Mastering
Household Electrical
Wiring
Second Edition

James L. Kittle

TAB BOOKS Inc.
Blue Ridge Summit, PA

SECOND EDITION
FOURTH PRINTING

Copyright © 1988 by TAB BOOKS Inc.
First edition copyright © 1983 by TAB BOOKS Inc.
Printed in the United States of America

Library of Congress Cataloging in Publication Data

Kittle, James L., 1913-
Mastering household electrical wiring.

Includes index.
1. Electric wiring, Interior. I. Title.
TK3285.K57   1988   621.319′24   88-5472
ISBN 0-8306-0987-3
ISBN 0-8306-2987-4 (pbk.)

TAB BOOKS Inc. offers software for
sale. For information and a catalog,
please contact TAB Software Department,
Blue Ridge Summit, PA 17294-0850.

Questions regarding the content of this book
should be addressed to:

Reader Inquiry Branch
TAB BOOKS Inc.
Blue Ridge Summit, PA 17294-0214

Table of Contents

To Violet L. Stancell

Preface

WHILE NEW METHODS AND IMPROVED POWer tools are presently used, skills and good workmanship are still needed for the job. Some people call doing a good installation an art. And it is! Pride of workmanship, neatness, and straight-and-level alignment of tubing or flexible cable, such as Romex, does improve a job and can result in more ready approval by the electrical inspector,

All construction methods and materials used today are explained and illustrated. Better and more illustrations are used to help you do the procedures correctly. This will help take the mystery out of what to do in a given situation.

Electrical wiring *is* fun, and with proper instruction, is easy. True, there is manual labor, including climbing, crawling under the house and in the attic, and sawing and drilling holes in wood and metal. Take your time, work carefully, do it right, and check your work. You will save money by using your own labor and buying your own materials. In the end, you will have a properly installed and neat job. Good luck!

Preface to the First Edition

BOOKS ON HOW TO DO YOUR OWN WIRING ARE many and varied. However, the predominant fault with some of these how-to books is that they are written by excellent writers who know how to write, but their information is second hand.

Some books are a compilation of many experts' combined efforts all in one big book of do-it-yourself projects. Very few of these authors, writers, or researchers have crawled around in hot (or cold) attics or dug their way under houses without basements, all the while watching for snakes and other crawly things. How many have been apprentices for six long, hard months doing all the hard, dirty jobs no one else wanted? Try drilling through a 6-inch concrete wall using a ¾-inch star drill and a three-pound sledge. Or, have you ever threaded 2-inch rigid conduit by hand?

I have done every procedure, every method, used every material, and devised better, easier ways of doing things, but also suffered smashed fingers and banged my head. Construction work is hard, cold, hot, wet, and dangerous. Electrical installation is a science, an art that is very technical and exact. Wires must be connected correctly, labeled properly, and carefully tested before use. This book is designed to help you achieve those goals.

Acknowledgments

TO MENTION EVERY PERSON WHO HAS, IN some way, made this book a reality is impossible. The two owners of the electrical contracting company, whose names I have forgotten, were the best teachers a young apprentice could ever have. They instilled in me a concern for good workmanship that has ever since been my goal.

The construction foremen for whom I worked in other trades helped me. Electrical inspectors H.A. (Lucky) Swager, director of building & zoning, Zion, Illinois; A.C. Holliday, electrical inspector, Troy, Michigan; and especially Mark N. Shapiro, who collaborated with me on the technical aspects and interpretations of the National Electrical Code. Mr. Shapiro is Treasurer of the Reciprocal Electrical Council of the Greater Detroit area. Two of his functions are the formulation of certain amendments to the National Electrical Code and the reciprocal licensing of electricians so that they can work in any member city. Mr. Shapiro is chief electrical inspector for the city of Madison Heights, Michigan and in addition is the code instructor at the Michigan Electrical Inspectors' School.

The following companies have provided illustrations and text for this book. Their cooperation and courtesy are greatly appreciated.

Arrow Hart Division of the Crouse-Hinds Company, Hartford, CT
Benfield International Corp., Jack Benfield, inventor and copyright holder, Ft. Lauderdale, FL
Crouse-Hinds, Electrical Distribution Products, Charlottesville, VA
Commonwealth Edison, Chicago, IL
General Electric Co., Plainville, CT
Hawaii Electric Light Co. Hilo, HI
Honeywell Inc., Minnetonka, MN
Intermatic Inc., Spring Grove, IL
Kester Solder, Chicago, IL

Leviton Manufacturing Co., Little Neck, NY
National Fire Protection Assn. Inc., Quincy, MA
RACO Inc., South Bend, IN
GTE Sylvania Inc., Danvers, MA
City of Troy, Michigan, Jay N. Winslow, Electrical Inspector Supervisor
Woods Wire Products, Carmel, IN
City of Zion, Illinois, H. A. Swager, Director of Building and Zoning

I would also like to thank my sons: Rex, who gave excellent advice on my drawings; Steve, who advised me on the operation of my camera; and Blake, who took many of the photographs.

Joanna Rademacher of the Warren Public Library, who directed me to research material and helped me select books about writing for publication.

The frienship and help of TAB BOOKS, Inc. staff, especially Raymond A. Collins, Kim Tabor, and all the editorial department in completing and submitting my manuscript is greatly appreciated.

I am ever grateful to Violet Stancell for her help with the manuscript revision and the correction of punctuation. The new edition will be a better book because of her help.

Introduction

THE USE OF ELECTRICITY IN OUR HOMES, factories, and places of business is so much taken for granted that we forget that without it, everyday living and working as we know it would not exist. Electricity in the home is a necessity.

Anyone who is handy with tools and is used to following directions can do residential wiring correctly and safely by following this book and the rules of the 1987 edition of the National Electrical Code (NEC), published by the National Fire Protection Association, Batterymarch Park, Quincy, Massachusetts.

Your house wiring can be remodeled, added to, and modernized. This book will guide you. In addition, this book will save you money, because you eliminate labor costs by substituting your labor and you buy all materials at your cost with no contractor's mark-up added on.

Methods and materials described here have been found to be practical from many years' ex-

perience in the fields of electrical contracting and related building trades.

You will be guided through your present house wiring. Each item and device is described and its purpose explained. Many points and procedures are emphasized and clarified so it is clear that "this is the only way to do this procedure; do not attempt to do it any other way." This is not to scare you off. It is to ensure that you understand and follow the procedure. The purpose of this book is to train you so the wiring you have done will be approved by the local inspector. The inspector's word is law and you want the work to be approved on first inspection.

The inspector is firstly interested in safety, and secondly he or she must be satisfied that the work conforms to the requirements of the National Electrical Code. This book and the inspector's excellent free advice allow you to save money and learn a skill, which will always prove useful. Use common sense, be careful, watch what you are doing, and you will do fine.

First, carry a notebook and a good flashlight and make a thorough inspection of the wiring in your home. Go through this book and make yourself familiar with the various chapters. You could even take it with you on your inspection. Note everything that does not look right or seems worn, broken, or damaged. Many houses have wiring done by persons who are not careful or just do not work neatly or accurately. In other words, the work looks sloppy. This could mean standard procedures were not followed and perhaps some wires were improperly connected.

Most local governing bodies will issue a permit for you to do your own wiring. You will be asked to fill out and sign what is called a Homeowner's Permit to do electrical wiring. By signing the permit, you state that you will do all the work yourself and not contract the work out to others. Your family members may help, but you are still responsible for the work. Upon completion of the work, the inspector—after inspecting for violations—will either approve or disapprove. The inspector will explain to you what things must be corrected and how to correct them. Most inspectors are very happy to help you. For your own safety, they want you to apply for the permit so that they can help you to do the job correctly.

Caution: Do not do any electrical wiring without first obtaining the necessary permit, then call for the inspector to make an inspec-

tion of the completed installation. Another reason to have the work inspected is for insurance purposes. Any fire will be inspected by insurance adjusters. If it is suspected that there was no permit and no inspection, the insurance company might not pay your claim, even though the fire might have been some distance from any suspicious wiring. These fire insurance adjusters are very knowledgeable on standard construction methods and they can spot code violations readily. Be advised of these facts and act accordingly! Be safe! Do a good job, check your work, and be careful.

Note: In the United States and Canada, single-phase, 60-cycle ac power-line house voltages vary from one area to another. Voltages might even be varied during different times of the year. Generally, there are only two voltages available for standard residential use: 120 volts and 240 volts. In some areas of the United States, 208 volts ac is available. Do not confuse 208 volts with 240 volts.

From time to time, you will come across references to house voltage listed as 110 volts, 115 volts, 120 volts, 220 volts, 230 volts, or 240 volts. Any appliance rated at 110, 115, or 120 volts is a 120-volt appliance. Any appliance rated at 220 volts, 230 volts, or 240 volts is a 240-volt appliance.

Although all possible measures have been taken to ensure the accuracy of the material presented, neither the author nor TAB BOOKS Inc. is liable because of misinterpretation of directions, misapplication or typographical error.

Chapter 1

Inspecting Your Home

It is likely that your house was wired by a licensed electrical contractor, and generally they do good work. Still, it is wise to check everything. Previous owners might have made improper changes or additions. When checking your wiring, take with you a small notebook to record defects and list the items to be changed. Also list any additions or special improvements you would like to make.

Throughout this book, for convenience, I refer to nonmetallic cable as "Romex®," to armored cable as "BX," and to flexible conduit (having no wires inside) as "greenfield."

THE NATIONAL ELECTRICAL CODE

All wiring in the United States should be installed in accordance with the National Electrical Code (NEC). The NEC is, in most cases, made a part of city ordinances by adoption and is incorporated into and made a part of the municipal building code. Other governing bodies such as townships and counties, and even states, incorporate the NEC into their laws and ordinances.

While the NEC itself is not law, the above entities do incorporate the NEC into their laws and regulations. Each such entity also adds their special additions to the NEC. These additions also have the effect of law and *can* be enforced. The NEC requires only minimum conditions in the wiring methods. The authors of the NEC are interested principally in safeguarding people and property from hazards arising from the use of electricity. Sizing and layout of the system are left to the electrical contractor.

The code has been in existence for 80 years. It is updated and added to every three years as required by new technology.

THE METER AND POWER SUPPLY

Start at the meter, whether it's inside or outside of the building. Note the wires coming from the utility pole (service drop). Check their condition. On old homes, the connection to the building wall might be porcelain insulators. Modern practice employs a service bolt and a tension sleeve to hold the three-wire service drop. This connection must be securely fastened to the building frame because of the heavy strain when wires are covered with ice.

The wires coming down to the meter or into the house can be either in conduit or service-entrance cable (SEC). If your house is quite old, there might be only two wires coming from the pole. Consider increasing the electrical ampacity to 100A as your first project. See that the conduit or cable is fastened securely to the wall. If the meter is inside, which is likely if there are only two wires, you will want to move it outside when you increase the ampacity. Figure 1-1 shows meters on a condominium. Where conduit or cable goes through the house wall, it must be made watertight. Water can travel through an opening and seep into electrical panels. Caulk this area securely.

ALUMINUM WIRING AND ITS HAZARDS

Some years ago, aluminum wire was used for interior wiring in houses. This caused some fires and destroyed many electrical devices such as switches and receptacles. This damage was caused by the nature of aluminum, which tends to flow and deform under pressure and heat. This problem was compounded by electrolysis between the aluminum wire and steel or brass terminal screws on the devices.

Many houses were rewired with copper wire and new devices and related parts were used. There are devices on the market marked "CO/ALR" that have *brass* screws. These

Fig. 1-1. Meter serving a four-unit condominium. Meters on the right are for the electric water heaters. Left meters are for power. Upper left meter is for outside security lighting. The center large box is the disconnect for the building and the cable extending from the bottom is the ground wire and connects to a cold-water pipe.

devices used with aluminum wire or copper-clad aluminum wire do not cause problems. Figure 1-2 shows aluminum entrance cable used presently.

I do not recommend the use of aluminum or copper-clad aluminum wire for 15A and 20A branch circuits. It is brittle and does not work well. Aluminum wire is used with good results for the service drop and the entrance cable to feed the distribution panel and entrance panel inside.

Fig. 1-2. Service-entrance cable. The lower photo shows the neutral, made of numerous fine wires, twisted to form a wire.

THE SERVICE-ENTRANCE PANEL

You will find the service-entrance panel either in the basement or utility room and sometimes in the kitchen. (In my parents' home, it was in an upstairs bedroom.) Note the condition of the panel and also the type. There might be just a main disconnect (switch), or it might be combined with the fuse or circuit breaker panel. The front might be hinged or attached with four or more screws. Turn off the main disconnect or trip the main circuit breaker. The panel might consist of one or more pull-out fuse blocks. One will be marked MAIN. Pull this one out; touch only the handle. Now, carefully remove the front of the panel. See Fig. 1-4, which shows the wiring found in the circuit box. At

the top you will see the feed from the meter. *These wires are still live—be careful.* As the service supplies 240 volts, there will be three wires, black, red, and *bare.* If conduit supplies the service, the wires might be black, red and *white,* or two black and one white. The white wire might actually be black with white tape or paint near each end. This is the neutral wire. Figure 1-3 illustrates a circuit breaker panel.

Take a screwdriver with a ⅜-inch wide blade and a good plastic handle (no metal in han-

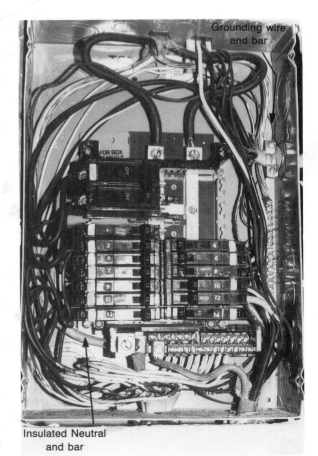

Fig. 1-3. Circuit breaker panel; the main breaker is outside. All bare *wires connect to the* grounding *bar on the right side of the panel.*

Fig. 1-4. A fused entrance panel. On the far right is the neutral bar; a number of white wires are connected. The pull-out blocks at the bottom are: left, air conditioning; right, electric range. The main pull-out is at the top right. Notice the main lugs just above the main pull-out. The neutral is at the bottom center, below the pull-outs.

dle) and tighten all screws having a wire connected to them.

CAUTION: Always follow the safety rule of using only one hand when working around and near live (hot, and they are hot) wires. Put your other hand in your pocket, because you might unintentionally reach for some metal object and receive a shock or worse. It is also a good idea to stand on a piece of dry wood when working near live wires. The screws on the terminals at the top of the panel will be *hot* so be doubly cautious if you tighten them. The neutral (white or bare) wire might extend down to the bottom or side before being connected to the neutral bar. Figure 1-4 shows a panel with type "S" fuses (time delay).

OUTDOOR LIGHTING

In the interest of security and to provide yard lights for convenience after dark, you might want to install outdoor lighting. Perhaps you have lights at the front and rear doors and the garage. If you want, this lighting can be increased to provide lighting around the perimeter of the house and garage.

To control such lighting, either a timer switch or a photocell can be installed. For decorative lighting during holidays, a number of outlets can be installed and can also be controlled the same way.

CHECKING THE BASEMENT

The basement and attic are the easiest places to check because the wiring is exposed and problems can be seen and noted. The distribution panel has many Romex cables leaving the top and sides. These will be either white, gray, or black. This is the sheath color; it is not the color of the individual wires. Each cable feeds a lighting or power circuit. Inspect and tighten all cable clamps holding the Romex or BX (metal-sheathed cable) to boxes or the panel. Tighten both the cable-clamp screws and the locknuts inside the cabinet or box. Some panels are located outdoors and have just the main disconnect breaker there. In some areas, even the distribution panel is outdoors. See that wiring on the basement ceiling is run neatly and is supported and anchored properly every 4½ feet.

Now turn on the main breaker or replace the fuse block. Leave the cover off for now or replace it if there are small children around. Trip one breaker off or remove one fuse. Now go through the house and look for lights that don't work. A fine device for testing outlets is a lamp socket with prongs on the end to plug into an outlet, like a plug on a lamp cord. This device is available at hardware stores.

When you find the dead (no voltage) circuit at an outlet, remove the plate and the top and bottom screws holding the outlet in place. Check for current again. *Be safe.* See if the screws holding the wires are tight; if not, tighten them. Also determine if the wire is aluminum. This is important. It is easy to check inside the breaker panel for aluminum wire. If any wire is aluminum, examine the outlet itself. If it is marked "CU/AL," the aluminum wire to it won't be acceptable. If it shows nothing, then there will be problems. Aluminum wire is white in color or it might have a copper coating on it (copper clad). This is somewhat better, but avoid aluminum wiring entirely if possible. Look at the end of the wire. Copper-clad wire is like plating and the wire end should show white. All junction boxes must have covers on them.

CHECKING THE ATTIC

An attic inspection should be much simpler

than a basement inspection, because most likely only the original wiring is there. All junction boxes must have covers on them. There also must be no wire connections made that are not made inside junction boxes. This is mandatory. Every wiring connection must be made inside a junction box or other enclosure, such as a switch or outlet box. All Romex within 6 feet of the entrance to the attic access hole must be protected either through bored holes in joists and covered with flooring or protected with running strips along each side of the cable. Use 1-inch-by-2-inch wood strips.

If you can see the ceiling outlet boxes for the downstairs ceiling fixtures, determine if they are solidly supported. Heavy ceiling fixtures over 10 pounds must be supported from the center bottom of the junction box. Do not support them from a strap screwed to the small box ears.

Ceiling fans must have special supports attached to the ceiling joists in the attic. These are now available at electrical suppliers and home centers.

GROUNDING

The main service entrance panel must have a ground wire leading to a metal cold-water pipe. The pipe must be in the ground for at least 10 feet to be considered a sufficient ground. The ground wire must be attached to the water pipe on the *street* side of the water meter. A pipe that is part of and connected to a municipal water supply is an excellent ground. If the grounding wire cannot be connected on the street side of the meter, then a bonding jumper must be installed to maintain a ground in the event the meter is removed from the premises. See Fig. 1-5 for the correct method. One house that I purchased had no grounding wire at all. Another

had the ground clamp left very loose. Modern wiring, using Romex or BX with a ground as part of the cable, assures safety to persons and also helps prevent electrical fires.

DETERMINING THE CAPACITY OF THE SYSTEM

Modern dwellings using many appliances require much more capacity than was required even 10 years ago. A minimum capacity of 100 amps is now required by the code. Even 150 amps is now common. The capacity of the service-entrance equipment is shown in the inside cover of the cabinet. The panel label will state "Underwriters Laboratories, Inc. 'listed' " and the amps capacity, such as "120/240 volts, 125 amps."

THE LIGHTING AND POWER CIRCUITS

Lighting circuits are supplied with 120 volts. They are protected either by 15-amp fuses or 15-amp circuit breakers if the wire is No. 14(14-2 w/gd.). If the wire is No. 12, it will be protected by either 20-amp fuses or breakers. Power circuits are supplied by 240 volts and they are protected by circuit breakers only, with a rating depending on their use, 30A up to 60A. The wire size used depends upon the appliance and corresponds to the rating of the circuit breaker.

REPAIRS

Items needing repair should be taken care of first so that the condition existing can be remedied as soon as possible. Faulty switches and worn electrical outlets are replaced easily. Everything is detailed fully in later chapters. Re-

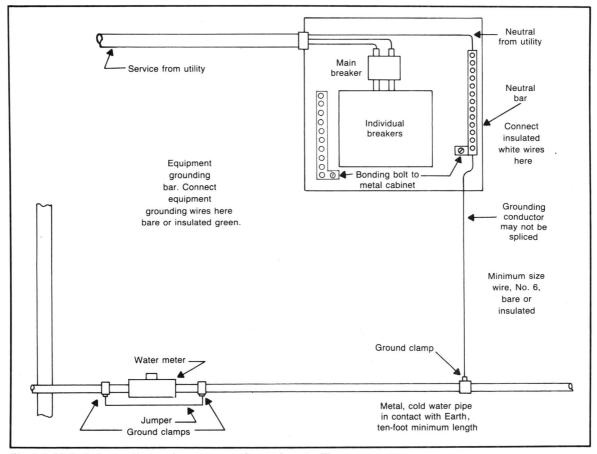

Fig. 1-5. Method of grounding service-entrance equipment in a dwelling.

pair the cords on appliances and lamps by just replacing the attachment plug or by replacing the complete cord and plug. Do the same with extension cords. If the cord itself is damaged, it should be replaced. If the damage is a bad cut at or near the center, the cord can be cut and two new connectors can be put on the cut ends, making two cords.

ADDITIONS

Any number of additions can be suggested for the person who is ambitious. Some sugges-tions for additions are to:

▸ Add a timer or photocell to control outdoor lighting.
▸ Wire a detached garage for lighting and power.
▸ Wire a workshop in the basement.
▸ Install a garage door opener. Include a pi-lot light to show that the door is open.
▸ Add another pilot light to show that the ga-rage light is on.
▸ Consider an emergency power system if your area has frequent power outages.

OUTLETS, SWITCHES, AND RELATED DEVICES

Basic wiring devices consist of many varied types, styles, and uses. The basic item would be the socket used for lighting and signaling purposes in lamps and fixtures. Next are the switches for controlling lighting and various motors. Also needed are the outlets (receptacles) for all the uses we put them to.

WORK TO BE DONE

By now you should have a list of work that needs to be taken care of immediately (such as defects and worn devices). Next, you should list work that needs to be done during warm weather, such as outdoor work. Winter jobs might be basement work, such as in a recreation room or wiring your shop for power equipment.

NOTE: Bear in mind that any major additions to the present wiring might overload the service entrance equipment; that equipment then must have its capacity increased before adding an extra load.

MATERIAL SOURCES

Wiring materials can be obtained from many sources including discount stores, hardware stores, retail stores such as Montgomery Ward and Sears, and their mail-order service departments. Certain devices such as range outlets, heavy service-entrance cable, greenfield cable, extra-large junction boxes, and service-entrance equipment can be obtained from large home improvement centers.

THE ELECTRICAL INSPECTOR

Municipalities having ordinances regulating electrical installations within their borders have an electrical inspection department as part of their building inspection department. The department requires that you apply for a homeowner's electrical permit. Talk with the inspector for your area. If you explain your plans, the inspector will most likely assist you. He or she might advise you to buy smaller-size wire for wiring special jobs such as a water heater, clothes dryer, or air conditioner. Smaller wire sizes will save you money and if the inspector approves, that is fine. All inspectors are very knowledgeable about certain aspects of installations that many electricians are not familiar with. Ask them anything about wiring you don't know. All it costs you is the fee for the permit.

Chapter 2

Safety

This chapter might well be the most important one in this book. Safety is important in everything you do, but it is doubly important in the wiring of a home. Read these pages carefully and be guided by them.

THE NATURE OF ELECTRICITY

Electricity has been observed for many centuries. The practical use of electricity developed during the past hundred years has revolutionized our lives.

Electricity exists in two forms: *static* and *current*. When you walk across a rug and then touch metal, you sometimes feel a shock and see a small spark. That is static electricity. This form of electricity is harmless but unnerving. The shock you feel is actually the jab of the spark at your finger. To eliminate this, hold a coin in your fingers and touch the metal. There will be no shock (but there might be a spark).

Current electricity is an entirely different matter. This form of electricity has been portrayed as an angry wolf with its fangs bared. This could be the case when people come in contact with live wires that appear to be installed correctly and safely. Anyone who has considered wiring to be safe and finds out differently by way of shock or worse should approach the original installer for redress.

Current electricity can KILL. Because of this ever present danger, all installations of electrical wiring must conform to the most recent National Electrical Code. Currently, this is the 1987 issue. If there is no local governing authority with jurisdiction, the wiring must still be done according to the NEC.

The National Electrical Code (NEC) specifically outlines the procedures by which electricity is controlled and made safe and useful to us without hazards and dangers. If after you have finished reading this book you decide to do your own wiring, keep these five Cs in mind: Care, Consideration, Checkout, Confidence, and Caution.

Care. In any electrical work you do, use care in planning the installation and in doing the actual work.

Consideration. Consider the work you are doing and its use by other people and how they come in contact with your installation. Make certain the devices and wiring materials are suitable for the use you are making of them. For instance, a dimmer switch for a fluorescent lamp installation might not work on an incandescent lamp installation—only some are interchangeable. Also make sure that wiring of all

devices is done according to recommended NEC rules. See Fig. 2-1.

Checkout. Check your work as you do it, after you have finished a certain section, and before going on. When the job is completely finished, again check and re-check all of your work.

Confidence. Have confidence in the work you intend to do and the work you have completed.

Caution. Remember to exercise caution at all times. Are there any bare wires exposed that will carry current? (This excludes equipment grounding wires). Are all connections tight? Are the correct wires attached to the correct terminals? Before starting work on elec-

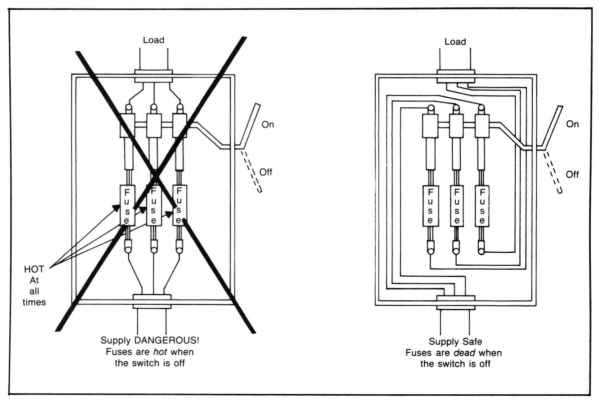

Fig. 2-1. An example of faulty wiring. In the left drawing, the fuses will be HOT, even though the switch is turned OFF. In the right drawing the fuses are dead as the wiring is correct.

trical equipment of any kind, has it been de-energized? Check at least twice for voltage. Check and double check all work you have done before energizing the installation.

SAFETY PROVISIONS OF THE NEC

The first statement the NEC makes is that its purpose is to safeguard persons and property from the hazards of electricity. The provisions contained in the NEC do not necessarily provide an efficient, convenient, or adequate installation. As stated above, the installation will be safe as required by the NEC, but not correct as to what the job will require for its proper use.

The NEC limits the amount of current allowed in various wires by specifying the American Wire Gauge (AWG) size and also the type of insulation for an allowable current carrying capacity (ampacity). The fuse or circuit breaker (breaker) at the distribution center has a specified rating. In the event the wire is overloaded, the fuse will blow or the breaker will trip to protect the wire from overheating and causing a fire.

Wiring methods and materials are also specified to eliminate substandard materials and poor installation and handling of those materials. A finished wiring job might appear to be perfectly installed even though it is faulty. This is the important reason to have the job inspected by the electrical inspector. The possibility of fire must always be considered. An electrical permit is the least expensive form of insurance you can buy.

AVOIDING SHOCK OR ELECTROCUTION

All wiring connected to a source of current must have a means to disconnect it from such a source, be it a private generating system or the public utility lines. Methods vary. There might be a lever-handle switch on the service-entrance cabinet, pull-out fuse blocks, or a circuit breaker. Any of these disconnect the wiring from any source of electricity. *Note:* The white or neutral wire must *not* be disconnected at any time, even when working on the wiring.

Before starting work on any wiring and after it has been disconnected, check for voltage with a tester. You might think you have tripped the correct breaker or removed the correct fuse for that particular circuit. Errors can occur. You might have picked the wrong circuit to turn off. Check again. Remember the ''C'' for caution. During my work as a heating serviceman, I had to replace many line-voltage (120V) devices. My final no-voltagecheck was to short the terminals of the old control to its case with a screwdriver. One time there was feed-back through another circuit and the screwdriver arced and blew the fuse. This way the tool rather than I received the shock. Always use caution. Sometimes the neutral and the hot (live) wires have been interchanged without anyone knowing about it. In this case, disconnecting the circuit would do no good and would do you harm.

I remember one instance where an oil burner motor was incorrectly wired. The switch for the burner disconnected the neutral wire instead of the hot wire, although it did stop the motor. On removing the motor from the burner, it started up in my hands because the wires were reversed. The hot wire was still connected all this time even though the switch was off. This illustrates why wiring must be properly done. Remember the five Cs of safety.

Another hazardous condition is shown in Fig. 2-1. A disconnect (switch) was wired so that it shut off current to the three-phase air conditioner it controlled. The hazardous condition existed because the wiring to the terminals of the disconnect was incorrect. The line (power supply) was connected to the bottom terminals. The conduit carrying the line voltage entered at the bottom. The load (wires to the equipment) was connected to the top terminals. This would disconnect the air conditioner, but it would leave the large fuses hot even though the switch was in the OFF position.

The right-side drawing in Fig. 2-1 shows the correct method of wiring this disconnect for safety. It is the approved method. Usually the top, always hot, terminals are covered with a fiber shield after wiring is completed. The hazardous condition encountered is that someone could turn off the disconnect and then grasp one of the large fuses to remove it and in so doing be grasping a live part. Be sure the equipment you are working on is dead so that you won't be dead.

WORKING WITH OR NEAR LIVE WIRES

If possible, do not work with or near live wires. In nearly all cases, the wiring can be disconnected before starting work. Many disconnects can be locked in the OFF position by a padlock to prevent accidental turn-on. Tag the switch to warn other people. As a last resort, it is possible to remove the electrical utility meter from its base. This disconnects everything.

As recommended in Chapter 1, use only one hand. Practice on a dead circuit until you can do this easily and with confidence. Touch only the part of the wire covered by insulation, not the bare end. Electrician's rubber gloves are available at electrical supply houses; do *not* use household rubber gloves.

Use only screwdrivers and nut drivers hav-

ing good plastic (not wood) handles. No metal can show at the top of the handle. Some screwdriver blades go all the way through the handle and form a head for pounding on the screwdriver. The plastic handle must be large enough so that you cannot touch the metal shank. The shank can be wrapped in electrical tape for more protection. When connecting two small wires by means of a wire nut, the hot wire should be held, by its insulated part, in your left hand. Bring the dead wire alongside and cap the two wires together with the skirt of the nut. Tighten the wire nut and the job is done. It is common practice to tape the skirt of the wirenut and continue onto the insulated wires for about one inch. This helps to secure the wirenut in place and covers any bare wires showing.

For obvious reasons, avoid handling live wires. For the very few times you might ever need to handle them, you might consider hiring a licensed electrician to make the final connection for you.

THE NEUTRAL AND GROUNDING WIRES

Power as supplied by the electric utility comes over three wires. Shown in Fig. 2-2 are two hot wires, R and B. The third wire is the neutral wire, N. Three wires are necessary to supply 240 volts for the range, electric water heater, and electric clothes dryer. If a central air conditioner is installed, this will also take 240 volts. The voltage between wires R and B is 240 volts. Between R and N it is 120 volts. Between B and N the voltage is also 120 volts. Circuits supplying 120 volts are used for lighting and portable appliances. The neutral wire must have white or natural gray color insulation. Wires R and B, the hot wires, may be black, red, or any other color *except* white,

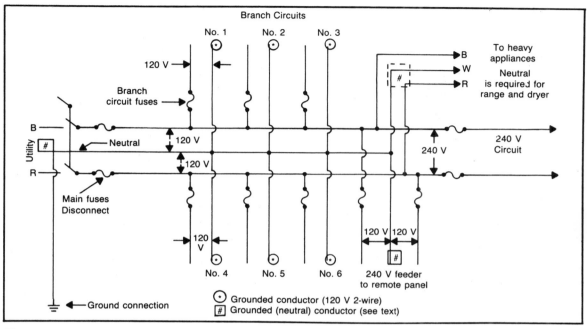

Fig. 2-2. Wiring layout for a residence.

green, or yellow. The neutral wire, N, has no voltage to ground and will not give you a shock. Service-entrance cable neutral is usually bare.

Large-size wires do not come in white or gray, but they can be identified as the neutral by white tape or paint at their ends where they are attached to the terminals. The neutral is never interrupted by a fuse or circuit breaker, switch, or other method. The only exception to this rule is where all wires, such as R, B, and N, are disconnected by a switch that cuts off all three simultaneously.

THE GROUND WIRE VERSUS THE GROUNDING WIRE

The *grounding electrode conductor* is the wire running from the service entrance equipment to the grounding electrode. This grounding electrode may be a cold water pipe, a driven

Fig. 2-3. Cut-away of a circuit breaker.

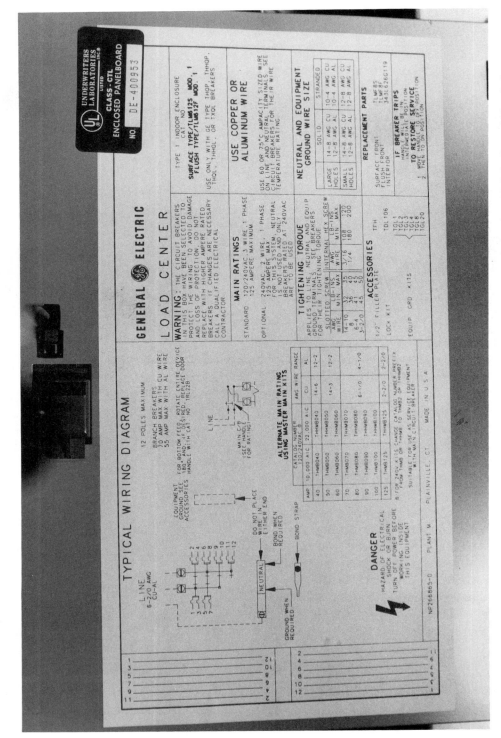

Fig. 2-4. Wiring diagram and information for a 125A load center. This also gives replacement parts and wire sizes. Made by General Electric Co.

ground rod or a buried metal plate. The ground wire is mechanically and electrically bonded to the grounding electrode by a ground clamp. An *equipment grounding conductor* is the wire connected to frames of motors, washers, and all metal noncurrent carrying parts of the electrical system. The designation on the Romex cable and the carton holding the cable of "14-2 with ground" refers to this grounding wire. This grounding wire never carries any current *except* in case of a fault or short circuit somewhere in the system. In this case, current would flow momentarily until the fuse blows or the breaker trips.

TESTING FOR HOT WIRES

Always test for voltage before doing any work on wiring. The only absolutely dead wire is when you can see both ends of the wire as well as the full length. First, test the circuit hot to test the tester. The tester might be defective and this test will show if it is defective. Now kill the circuit and test again. This time the tester should not register. Try the tester on another live circuit and then on the one to be worked on. *Be sure to tag the disconnect and lock it if possible. Be absolutely sure the circuit is dead before starting work.*

Figure 2-3 shows a cut-away of a circuit breaker. The ground fault circuit interrupter GFCI in Fig. 5-14 should be in every place in the house where a person could receive a shock due to wet conditions (a bathroom) or from contacting a metal object.

Article 210-8(a) of the NEC requires GFCI's to be installed in the following locations: bathrooms, garages (except in inaccessible places, like garage door operator outlets), dedicated space (that which is not needed, as behind a refrigerator), outdoors where there is direct grade access to the house, basements (for power tools, new in the 1987 NEC), within 6 feet of kitchen sinks, and in boathouses.

See Fig. 2-4 for a copy of a load center chart and wiring diagram.

Chapter 3

Tools

Tools are very important in working with electricity. Look for top quality and good design when you are buying your tools. Top-quality tools, if properly used and cared for, last a lifetime. The cheap ones will not. Some uncommon tools you might need are illustrated in Figs. 3-1 through 3-5.

PLIERS

I prefer the nationally advertised brands and especially those made in the United States. These brands include Lufkin, Stanley, Crescent, Craftsman (excellent guarantee), Channellock, Klien, Proto, Snap-On, Blue Point, and others. For certain hand tools I choose specific brands. Channellock groove joint pliers are of superior design even though others look identical. For electricians' linemans' pliers (sidecutters) and diagonals, Klien's are the best. This brand is usually only available at electrical wholesalers. They will readily sell you tools. Also ask about

buying wiring materials there. I have a pair of Crescent diagonal cutters at least 30 years old. They work like new and will cut bolts up to $\frac{3}{16}$ of an inch in diameter. Never cut anything but copper wire with your lineman's side cutters (pliers).

SCREWDRIVERS

Screwdrivers are very important tools. To tighten screws properly, the screwdriver blade must fit the screw slot snugly. Too large or too small a blade will damage the slot and perhaps ruin the screw. For most terminal screws, use a $\frac{3}{8}$-inch-wide blade that has a blade thickness to properly fit the screw slot. This is very important: Be sure the handle is made of solid plastic with no metal cap on the end or metal shaft through the handle. Some such as Crescent have a handy rubber sleeve over the plastic handle for gripping.

Screwdrivers for small screws need to be

16

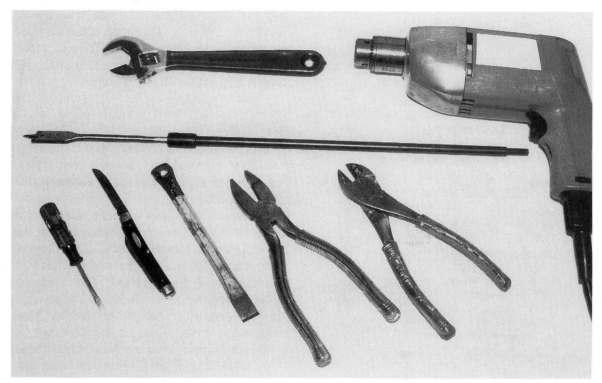

Fig. 3-1. Top: 8-inch adjustable wrench, ⅜-inch electric drill. Center: ¹³⁄₁₆-inch spade wood bit and bit extension. Bottom: pocket screwdriver, pocket knife, cold chisel, electrician's side cutting pliers, electrician's diagonal cutters (which cut to the end of tool).

smaller in width and thickness to match the screw slots closely. Three sizes will do fine. For Phillips-type screwdrivers, two or even one medium size may be all that's needed. These types of screws are not common on electrical equipment. Always use a screwdriver with a plastic handle.

A very popular type of screw-holding screwdriver called the Quick-Wedge is used by electricians when working on live terminals. The screw can be started in the hole and tightened, then the screwdriver can be removed and you never touch any live parts.

NUT DRIVERS

Although nut drivers appear similar to screwdrivers, the business end is a socket, as in a socket set—except with a screwdriver handle. Some are part of a ¼-inch socket set that

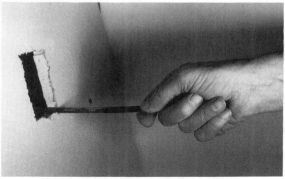

Fig. 3-2. A hacksaw blade used to cut an opening in drywall for a wall box. The handle is taped for hand protection.

17

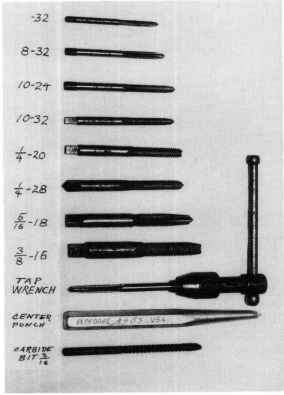

Fig. 3-3. *Different size taps used in electrical work. The tap wrench will hold all the taps shown. The center punch is used to make a center for starting the drill. The carbide bit drills holes in masonry.*

includes a ratchet handle. Others are individually complete with only one size socket on a handle. This type has a hollow handle to accommodate a long bolt that must have the nut run down quite a distance. Nuts are common on home appliances, service-entrance equipment, and distribution panels. One-quarter-inch drive-socket sets retail for about $10 to $15. Individual nut drivers cost from $1 to $2 each.

WRENCHES

Hex head nuts, bolts, and sheet-metal screws are tightened either with nut drivers,

open end, or adjustable wrenches. This is with dead equipment only. The larger sizes are used for types of conduit fittings (among other things). Sizes to buy are 6-inch, 8-inch, and 10-inch adjustable. Buy top quality.

HACKSAWS

The best tool for cutting metal is the hacksaw. They are made in many different styles. The common style is the frame hacksaw that takes a ½-inch by 10- or 12-inch-long blade. Blades have three sizes of teeth with spacing of 18, 24, and 32 teeth to the inch. Buy the 32-tooth saw blade in high-speed steel. The cost is about $1 each, but they will last much longer than the less expensive ones. The teeth are hardened and the back is flexible. The less expensive ones will break in the middle. Most hacksaw frames are adjustable to take either 10- or 12-inch blades. Buy a medium-grade frame. There is also a small keyhole-type hacksaw useful for cutting drywall openings and for use in restricted areas. Another type of frame uses broken blades as the keyhole type uses its blade.

METAL SNIPS

Although most metal can be cut with the hacksaw, thin sheet metal must be cut with tin snips. Metal snips should be chosen with care. The best brand of snips are made by Wiss. They make every kind of hand shearing tool from heavy sheet-metal snips down to embroidery scissors. Aviation snips are not absolutely necessary, but they are extremely useful. If you have much sheet metalwork, "cuts right" and "cuts left" models are really necessary. Years ago I purchased a "non-Wiss" brand and I have been having trouble making them work right for

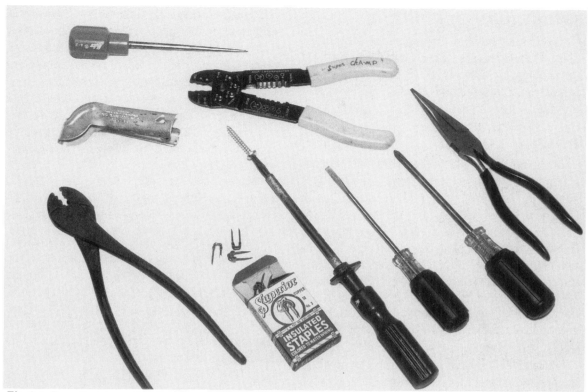

Fig. 3-4. From the top: scratch-all, universal tool for crimping wire terminal ends on wires, cutting wires, skinning insulation from wire ends. Bottom row: old style crimper, insulated staples for low voltage circuits, screw holding screwdriver, blade and Phillips type screwdrivers, long nose pliers (needlenose).

me ever since. Keep in mind that no one brand is best in design for all tools.

BOLT TAPS AND DIES

All cabinets, wall boxes, and junction boxes have tapped holes for mounting screws, as do many devices. These items usually have good threads tapped in them, but defects show up at times. Wall boxes already mounted in place can be saved if they have bad threads by using the correct size tap. Taps cost from 75¢ to over $2 each. All that are needed are 6-32, 8-32 and 10-32. Sometimes the 10-24 thread is needed. This is the common 3/16-inch bolt size. A tap wrench handle holds the tap so it enters the hole

straight. These handles cost between $2 and $3. Dies with the accompanying die stock (handle) are used to thread bolts and machine screws. You will find occasional uses for them.

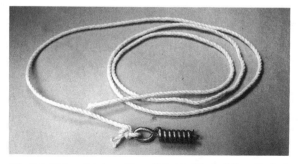

Fig. 3-5. "Mouse" (wire solder coiled up) with chalkline attached. Used for fishing cable inside hollow walls.

LOCKING (VISE) PLIERS

The very popular locking pliers are a great help. With no vise at hand, parts can be held for drilling holes safely. Vise-Grip® is the best brand. It is used almost exclusively by industry and is very reliable.

POWER TOOLS

The best time and labor savers are the portable power tools used in the home, building construction, repair shops, and industry. As in electric drill motors and hand-held power saws, price is an indicator of quality. The amperage listed on the name plate also is another indication of good quality. When buying an electric drill motor, select Black and Decker®, Skilsaw®, or Rockwell® brand name motors. A ⅜-inch capacity reversible variable-speed drill motor is a good investment. Buy in the range of $40 and you might never need to buy another.

Electrically operated power tools must be maintained in first-class condition. Older models might have three wire cords with three-prong grounding plugs on them. They are supposed to have the metal housing connected to the grounding (green) wire that is in turn connected to the grounding prong of the plug (U shape). You can determine if this is the case by using a continuity tester. Put the probes of the tester (one on the metal housing and one on the U blade of the plug). A light indicates that the case is grounded properly. Now put one prod on the case and the other on one of the flat blades of the plug, then place it on the other flat blade. A light in either of these last two tests indicates a ground to the housing. Dangerous! This defect must be repaired immediately before using the tool again.

A much older tool has only a two-blade plug. This is not to be confused with the modern *double insulated* tools sold today. These new tools come with a two-prong plug and have a very good record of safety. Buy them with confidence. Three-prong grounded power tools are still available for purchase, but they are generally available only in heavier usage types. The only way to be safe with this old two-wire tool is to rewire with three-wire cord and plug and ground the frame with the green grounding wire in the cord.

HAND TOOLS WITH PLASTIC-COATED HANDLES

Many hand tools such as pliers, wrenches, and tin snips are sold with plastic grips on the handles. This plastic is put on by a dipping process. These plastic grips are not to be considered as having any insulating value at all. They are merely a comfort grip. They will not provide insulation from electricity. This information is on the authority of the local electric utility (Detroit Edison). This is in the form of a notice posted in their repair shops for the information of their employees and others. They warn against using and depending on this so-called insulation. The only true source of insulated grips is the electrical wholesaler. They sell them for self installation. The grips are softened in boiling water and then forced on the handles by tapping with a block of wood while still hot. They are then sold as insulating grips. Their thickness is at least three times the thickness of the dipped grips and the plastic is a different material. In any case, use caution around live wires.

DRILL BITS

As with other tools, drill bits come in different grades and qualities. For drilling steel, buy only high-speed steel bits. These can be used to drill all other materials except masonry. Carbide-tipped drill bits are the only drill bits that can be used for drilling all forms of masonry. The small sizes such as ³⁄₁₆ inch, ¼ inch, ⁵⁄₁₆ inch, and ½ inch will be needed. Smaller sizes are for plastic inserts used with No. 10 sheet-metal screws to mount metal boxes on masonry. The ½-inch drill will accommodate ¼-20 lead anchors using plain ¼-inch bolts for heavier items. They will also accommodate the insertion of toggle bolts in hollow tile and concrete blocks.

If you need to drill larger holes for conduit or cable, larger and longer carbide drills can be rented at rental businesses. Sometimes hammer drills are needed and they can also be rented there. Long lengths are up to 24 inches and larger-diameter drills are also available.

ELECTRICAL TESTERS

Testers that are used on live circuits are used to test for the presence of voltage (power) in equipment in use or having unidentified problems. Presence or absence of power is determined and is corrected or defective components are replaced. Electricians and service technicians use voltage meters constantly.

The commercial heavy-duty testers are on the market for about $20. They have long life and some have replacement test leads available. A pocket tester costing under $2 is available in hardware stores. This tester can be used for testing circuits that are live (hot).

The *continuity tester* or resistance meter is not for use on live circuits! This tester is used on nonenergized circuits and certain controls to determine what problems are causing the trouble in equipment or wiring. One continuity tester is built into a flashlight. Brite Star manufactures such a tester. The flashlight has a radio jack in

the end cap and is supplied with two test leads having a radio plug on one end that is plugged into the end cap. The other ends of the leads have test prongs. By turning on the flashlight switch, the tester is ready for use. The light will not go on even though the switch is pushed to ON. Touching the two test leads together or testing for continuity will cause the light to go on. Removing the plug will also cause the light to go on. Turn the switch off when done using the tester. All continuity testers must have a current source (batteries).

Electricians use a homemade tester for testing wiring in an unfinished building that does not have the wiring connected to a source of power. In this way, all wiring is checked out before power is applied. This eliminates any defects that would show up as blown fuses or worse and might damage expensive equipment. This tester is made from a 6-volt lantern battery and buzzer or bell and long test leads.

A second type of tester is a vest pocket flashlight with one test lead and the case as the other lead. Another type is a screwdriver with a transparent handle enclosing a lamp bulb. The screwdriver is one lead. Another lead about 12 inches long provides a handy tester. The pocket flashlight seems to be less handy as only one lead makes it more awkward. The case is used as the other lead. These cost $4.

The type most used by electricians when testing new installations is homemade. A 6-volt lantern battery and a door buzzer plus leads are assembled by taping the buzzer to the battery and having 2-foot leads to handle greater spacing of the parts to be tested. The battery and buzzer might cost $6. The battery will last a long time, but it is not as portable or handy as the Brite Star flashlight.

As electronic buffs usually have a volt-ohmmeter in their tool kit, this can also be used.

It can be used as a voltage tester or as a continuity tester, the latter by using the ohmmeter scale. This tester is also awkward. I find I have to watch the needle while holding the prods on the wires or terminals being tested. With the flashlight type, the light will be seen out of the corner of the eye without taking the eye off the test area. This seems like an easier test procedure.

The pocket tester has a neon bulb and short leads. The whole tester is the size of a large pencil. This tester has been known to give false readings because of the neon bulb and the fact that a feedback can occur from other live circuits. Even though it is very low in cost, I do not recommend it.

EXTENSION CORDS

With the heavy usage that extension cords get, it is advisable to buy as good a cord as you can afford. A 50-foot, heavy-duty cord (16-3 or 14-3) should cost perhaps $15. The heavier wire is to prevent voltage drop on long runs. If you have all the current shut off, you need this long length. The three wires are the black, white, and green. The green is for grounding. Do not buy two-wire cords.

Some extension cords have a trouble light on the end of the cord and others do not. If the trouble light is a good one it will have an outlet or maybe two on the light handle. If the cord is three wire (with ground wire), it should have a three-slot grounded outlet. Check the lamp guard. It should be easy to change the bulb without having to dismantle the guard.

When you buy a bulb or bulbs for the trouble light, be sure to get the *rough-service* designated bulb. These bulbs cost more but last a very long time despite rough treatment. I have

dropped them in the guard, and they have continued to function properly.

CAUTIONS REGARDING POWER TOOLS

Power tools are great labor savers, but they must be used with exceptional care. I have watched many workmen, amateurs and professionals, injure themselves or damage the item they were working on due to careless handling. I have also been injured or damaged something while using tools.

Drill motors are usually responsible for problems rather than a saw. The saw seems to command more respect in this regard. Large holes drilled through the back or side of a junction box should be started by using a ⅛-inch pilot drill, then a ¼-inch drill, and finally a ⅜-inch or ½-inch drill. When a large drill breaks through the metal, it has a tendency to "hang up" (grab into the material). This can very easily spin an item like a 4-inch junction box on the end of the drill letting it bang into your ankles.

To hold such an item, drive two nails through the holes in the back of the box into a length of 2 × 4. Stand on the 2 × 4 while drilling.

Important: When drilling anything overhead, be sure to wear safety goggles. Eye protection is vitally important to preserve your eyesight. Safety regulations now recommend goggles at all times when working with tools. At least wear them whenever there is any danger of foreign material being thrown on dropped in your eyes. If you do get something in your eye, go immediately to the emergency room of the nearest hospital to have it removed. This is especially needed in case of metal in your eye. Take no chances.

TOOLS FOR CONDUIT WORK

Tools for conduit work are different in some ways from those used with Romex and BX. To bend conduit, it is necessary to purchase or rent a Thinwall bender. Thinwall tubing, which is the only type you need to use, must be bent with this special bender to prevent crushing the tubing. The bender costs about $14. Newer models of benders are able to bend both ½-inch and ¾-inch Thinwall. They also cost more. Benders come with directions in the form of an illustrated booklet. You might have to practice before becoming good at it.

The regular 90-degree bend is made with the bender hooked onto the conduit, in the right place, and pulled down by the handle. If the next bend is in the opposite direction, the bender is set on the floor, handle end down, and the conduit is inserted in the bender head which is up in the air. Then the conduit itself is used as the handle to bend itself down to the right angle. Your foot is on the handle end of the bender resting on the floor to keep it from slipping. This maneuver is tricky but it works. It took me quite a while to get the knack of it. Refer to Figs. 3-6 through 3-16.

A device known as a "hickey" is used for bending rigid conduit (similar to water pipe). Never use water pipe for conduit. Rigid conduit is difficult to work with because it requires pipe dies, a pipe vise, and other related equipment. You will never need to use rigid conduit.

HOLE SAWS

If it is necessary to make large holes in metal, hole saws are very useful. Metal cutting saws are expensive but necessary to bore holes in metal. Uses would be for cutting holes in cabinets to insert conduit or Romex where there is no knockout available. You will find other uses for them. Buy only high-speed saws.

TIPS ON TOOLS

In many cases, you can purchase good used hand tools and possibly power tools at garage

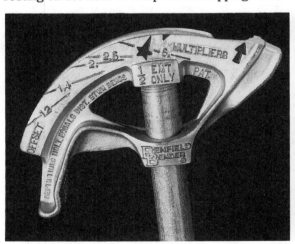

Fig. 3-6. Front view of Thinwall (EMT) bender manufactured by Benfield Co. (Courtesy Jack Benfield, copyright owner and inventor).

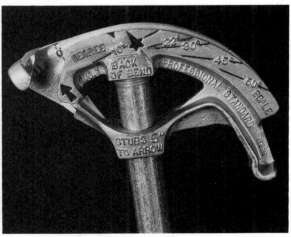

Fig. 3-7. Rear view of Benfield Bender for EMT. (Courtesy of Jack Benfield, copyright owner and inventor).

Fig. 3-8. Bender is set on the Thinwall to make the first bend of an offset for a handy box. Notice that the bender hook is not at the extreme end of the tube. This prevents crushing the tube end.

Fig. 3-9. First bend has been made. Bender handle is at 45°.

Fig. 3-10. Second bend is being made in the air. If a floor edge or similar flat surface is available, the bend can be made on the floor with the first half of offset over the edge.

Fig. 3-11. Second 45° bend completed. Note: Both bends must be in the same plane or bend will be lopsided. When making the second bend, align the first bend with the bender handle.

sales and second-hand stores. Power tools should be inspected carefully before purchase to determine their condition. You might be able to find good bargains this way. Just be cautious because they might not be returnable.

Figures 3-17 through 3-25 show some tools and techniques for plugs, connectors, and testers.

Fig. 3-12. The completed offset. Thinwall will now fit handybox.

Fig. 3-13. Shows a wall box installed using Madison supports. The switch controls a ceiling fixture. In this situation, the neutral wire is in the box and a combination switch and receptacle can be installed.

Fig. 3-14. An offset made in Thinwall to allow it to lie flat against the wall.

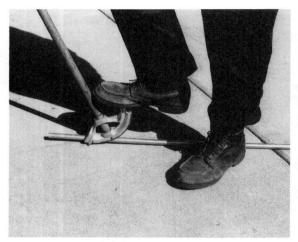

Fig. 3-15. Start of a "stub-up" 90° bend. The required hight of the stub-up is less than 5 inches; the bender mark is set at this point.

Fig. 3-16. Finishing the 90° bend. Height is correct.

Fig. 3-17. Homemade 120/240V tester. The sockets are wired in series and 120V bulbs are used. Full brilliance means 240V, half brilliance means 120V.

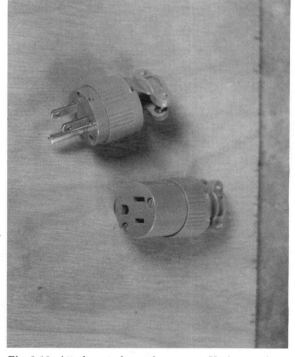

Fig. 3-18. Attachment plug and connector. Used to repair or make new extension or power tool cords.

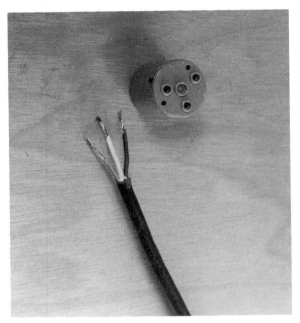

Fig. 3-19. Plug ready for attachment of cord wires.

Fig. 3-21. Connector ready to be assembled.

Fig. 3-20. Cover for connector shown in Fig. 3-19.

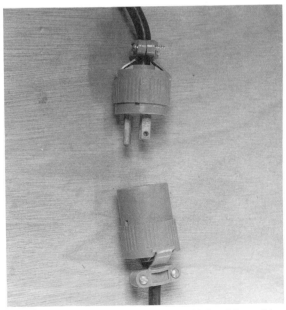

Fig. 3-22. Both ends are completely assembled and the cord has been repaired properly.

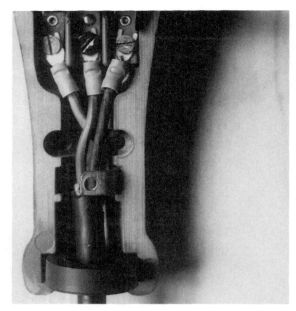

Fig. 3-23. A new cord has been attached to a trouble light. Notice the crimp-on terminals.

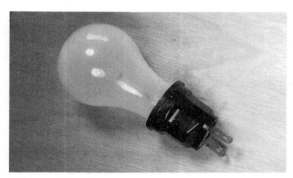

Fig. 3-24. Line voltage "plug-in" tester. A socket with prongs to be plugged into a receptacle or other similar outlet.

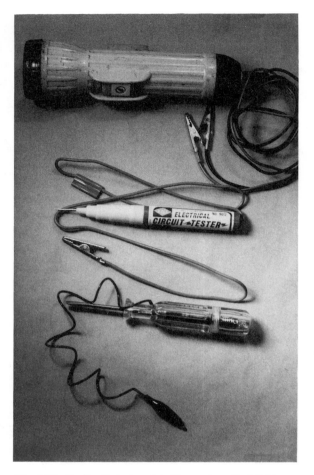

Fig. 3-25. Continuity testers, top: To use, turn flashlight on, then plug test cord into flashlight end cap. Flashlight will light only when test prods are touched together or to any terminals that are "made" (continuous). Sells for about $12. Lower testers are the same except test leads are permanently attached. Costs are: center, $3, lower, $5.

Chapter 4

Service-Entrance Equipment

Service-entrance equipment, as the name implies, is the equipment and necessary cable to bring electric service into a building. The meter and its connections, including the base, are included as components of the equipment. When the utility supplies service to this equipment, the wires might be overhead or underground depending on the utility's supply lines in the area. It also might be installed underground at the owner's option, even though the service to the area is overhead on poles. In this case, you have to pay for running the cable underground and up to the utility pole. But, you might be able to do this work yourself, paying only for the materials and trenching. Review Fig. 2-2 and other figures throughout this chapter.

The service-entrance equipment is intended as a means of disconnecting the electrical supply to the premises. It consists of a circuit breaker or fuses and a switch (called a fused disconnect). This may be either inside or outside the building. Figure 4-1 shows a fuse panel for an air conditioning unit. The circuit breaker is used in many houses although fuses are still one of the best overcurrent protection devices available. The breaker is convenient and it can be reset. The fuse is very dependable.

The utility will make the connection to your service-entrance equipment when the installation has been approved by the local electrical inspector and the utility. Before starting any work on your system, visit the local offices of your utility and city inspection department. Have with you a simple plan showing the proposed location of the service equipment in relation to the utility's lines. Figures 4-2 through 4-8 show various service-entrance panels.

POWER SUPPLY

Power to modern houses is called the three-wire 120/240V system. This system consists of three wires brought from the utility pole

Fig. 4-1. Fuse panel for air conditioner condensing unit.

Fig. 4-3. The 60A service entrance panel less wires. This provides a clear view of the pull-out fuse blocks.

Fig. 4-2. A mock-up showing a 60A service-entrance panel. Top left, service entrance cable from meter. Top right, service-entrance cable to the electric range. Top center is the connection for the line neutral and the load neutral to the electric range. A connection is also provided for the ground wire to a metal cold water pipe or other grounding method. A busbar extends from this neutral down behind the panel interior to secondary neutral bar for circuit white wires to be connected.

or underground to the house. If overhead, the wires are twisted together to form a sort of cable. There will be two insulated wires and one bare wire. This is called a *service drop*. The bare wire serves as a messenger cable supporting the other two and is usually galvanized steel (stranded). This is the grounded neutral and it is attached to the building by a tension clamp and bolt through the building framing. Underground lines from the utility are called service laterals (the wires are usually individual). This underground wire is called USE (underground service lateral). All three wires are insulated.

As shown in Fig. 2-2, the three wires pro-

Fig. 4-5. Close-up of terminals for the mains (center of picture). Neutral terminal is above right main terminal. The neutral bar is just above the neutral main; notice the white wires.

Fig. 4-4. Service-entrance panel, circuit-breaker type. The 240V mains connect to the two large terminal screws, one on each side of the panel above the breaker section. The neutral terminal is just above and to the right of the two hot terminals (mains). Double breakers (240V) are in the center section. Single breakers (120V) are below the double breakers.

Fig. 4-6. Close-up of the three double breakers. These protect the electric range, air conditioner, and a receptacle for an electric clothes dryer. Shown at bottom are single breakers side by side. Each breaker is then fed from each side of the 240V service, which equalizes the load properly.

vide either 120V or 240V, depending on the connections between the wires. The two outer wires between them provide 240V without a grounded wire (neutral); this is for water heaters and 240V motors. Either one of the outside wires and the neutral provide 120V for small appliances and lighting. For electric ranges and electric clothes dryers, the two outside wires plus the neutral (grounded) provide 120V or 240V. The utility lines that supply power to the area, usually from the rear of the property, consist of two separate sets. The upper (higher on the pole) lines carry the high voltages, as much as 2400V or more. This is the transmis-

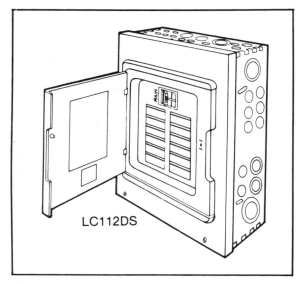

LC112DS

Fig. 4-7. Service-entrance panel, includes main breaker at top (Courtesy Electrical Distribution Products Division, Cooper Industries, Inc.).

sion line from the utility substation. See Figs. 4-9 through 4-11.

Because of the characteristics of alternating current, the voltage can be lowered or raised by what is known as a transformer. This transformer consists of many turns of wire wound around soft iron laminations (plates). There are two separate windings known as the *primary* (the voltage going into the transformer) and the *secondary* (the voltage leaving the transformer).

At intervals, transformers are mounted on poles. These reduce the voltage to that supplied to the houses, namely 120/240V. One transformer supplies power to a group of houses (two or more). These lower voltage wires are below those carrying the high voltage. Even lower on most electric utility poles are the telephone and cable TV lines. Thus, each different set of wires serves a different use and is always in its relative position on the utility pole. See Fig. 4-12.

There is a ratio between the number of turns of wire in the primary and secondary windings. This ratio may be 10:1 (a voltage-reducing transformer) or 1:10 (a voltage-increasing transformer). The voltage-reducing transformer is the one on the pole supplying power to your house. The high voltage from the higher (on the pole) wires, usually 2400V, is led into the primary connection of the transformer and reduced to one-tenth of the 2400V emerging from the secondary connection as 240V for use in your house. By making a center tap on the transformer, 120V is provided for lighting and small appliances.

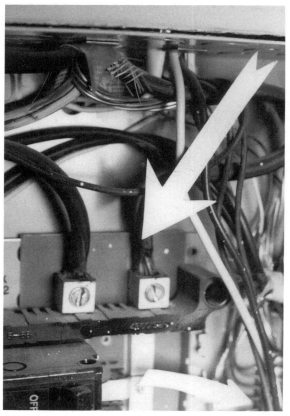

Fig. 4-8. Close-up of the service entrance cable lugs (terminals). (Arrow points to right lug.)

Fig. 4-9. Substation and distribution center. Detroit Edison Co.

Storms greatly affect overhead open wiring on poles and cause power outages and downed wires (especially in the winter). The practice now is to run utility lines underground. This method is more expensive, but it eliminates a great deal of work during severe storms.

Reading an electric meter is comparatively easy (see Fig. 4-13). Start from the left and always read the number the hand has *passed*. All the dials are connected by gears between them. The gears are so proportioned that the hand on any dial rotates 10 times faster than the one to the left of that dial. Notice that one hand will rotate counterclockwise while the next one will rotate clockwise. Notice that when the right dial hand moves just past ''0,'' the reading of the meter will be, for example, 1880. The right dial hand continues and next records the reading as 1881, and so on. This arrangement is similar to the odometer on an automobile where each number wheel makes one revolution to move the next wheel only one number.

You can keep track of your electric usage by reading your meter at intervals. Subtract the last reading from the present reading and you have the usage for that period. This is the usage in kilowatts (kW). The average family uses about 500 kW each month. If the water heater has a separate meter, you can monitor your hot water usage and perhaps use less hot water or lower the setting of the heater thermostat. See

Fig. 4-10. Modern high-tension tower (175,000V).

Fig. 4-11. A high-tension line dead-ended at this pole. Top wires are usually 4800V and these feed a shopping center. Voltage dropping transformer is on customer's property.

on. Try this setting of the thermostats for a week or two. You might be able to reduce the set point even lower to 125° F or less. Save money. Refer to Chapter 14 and Figs. 14-26 through 14-36 for directions on installing an electric water heater timer to save money when heating water with electricity.

Some utilities might still control electric water heaters by means of a timer set up to power the heater only at low demand loads at the

Fig. 4-14. Lowering the set point (temperature setting) is a simple procedure.

To do so, shut off the power to the electric water heater before doing anything to the heater. Standard heaters usually have two thermostats, one upper and one lower. The thermostat is attached to the actual heating element that is immersed in the water of the tank. If you are sure the power is off, remove one of the thermostat cover plates. (Figure 14-29 in Chapter 14 shows a thermostat and heater element exposed. The arrow points to the thermostat dial and pointer.) The pointer is factory set at 140° to 150° F. First try a set point of 130° F. Do the same to the other thermostat. Carefully replace the covers and turn the power back

Fig. 4-12. Utility transformer in residential area. Top wires are 4800V. Wires directly below are 120/240V to the homes. Notice the lightning arrestor just above the transformer.

Fig. 4-13. Electric watthour meter. Reads either 01879 or 00879. Meter reader would check last reading to see which was correct or report a defective unit.

Fig. 4-14. Water heater timer, property of utility. This shuts off the electric water heater during high usage periods.

wires, one bare and two insulated. The bare wire, the stronger of the three, is the neutral and is used to support the cable by means of an adjustable *tension connector*. The point of attachment to the building must be a minimum of 10 feet above grade (in some areas it's higher) and 10 feet above the sidewalk. Other clearances are 12 feet above a private drive and 18 feet above public highways. The utility maintains these clearances on its own wires. Clearances over roofs must be a minimum of 8 feet. The building where the final attachment is to

Fig. 4-15. Underground service lateral to a single home. The meter is outside; the main 100A circuit breaker and distribution panel are inside.

generating station. This means that the heater would be off for extended periods. With this arrangement, the heater size is understandably larger (80 gallons) and the water temperature is higher (150° F or more). The billing is a flat rate per month. If you run out of hot water, there is often another circuit that can be turned on, but it is metered through the regular meter at the high rate. This arrangement might still be used in some areas, but it is not popular now. Most electric water heaters now come with a 150° F setting that can be lowered.

SERVICE DROP
AND SERVICE LATERAL

The overhead lines from the pole are called the service drop. This consists of three twisted

Fig. 4-16. A meter assembly for five stores in a shopping mall. One store is larger. Service entrance equipment is inside each individual store.

be made must have this attachment point high enough to give this 8-foot clearance. There are two exceptions to this. The first is where clearances over roofs range from 3 feet minimum where the voltage between the conductors does not exceed 300 volts and the roof has a slope of not less than 4 inches in 12 inches. The other clearance exception is where a service head extends through the overhang portion of a roof. This clearance minimum is 18 inches at the point of attachment to the mast. Clearance around building openings must be maintained a minimum of 36 inches away from the bottom and sides

Fig. 4-17. Meter base for a single home. Underground line (service lateral) rises into meter base (line) and the service entrance cable goes into the basement to feed the breaker panel.

Fig. 4-18. Continuation of Fig. 4-21 shows service lateral into ground. (Telephone line is shown also.)

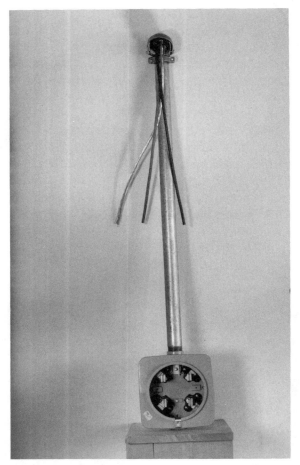

Fig. 4-19. A mock-up of a service mast and meter base (socket). An actual mast is much taller so the service drop has the proper clearance.

nection. Some utilities furnish and install the ser-service lateral, but you will be billed for this work. Raceways that may be used are duct, nonmetallic conduit and galvanized rigid conduit. Galvanizing must be approved by the inspector to guard against corrosion, otherwise, a coating that is noncorrosive must be applied to the conduit. Cable or three separate wires must enter or leave conduit whose ends are fitted with insulating bushings. This would be where they enter a building or leave the ground to enter a conduit that will go up a pole.

Because underground is classified as a wet location, the insulation must be one of the fol-

Fig. 4-20. Close-up of a meter socket. Note the three wires connected to the line terminals—black on the left and red on the right (both could be black). The neutral is bare or can have insulation of any color. If other than grey or white, the neutral must have white paint or tape applied to each end for identification.

of door and window openings. Areas above such openings have no minimum clearance, because they are considered out of reach.

The underground lines from the utility connection to the premises are called service laterals. See Figs. 4-15 through 4-18. They are direct burial cable or three separate conductors. Raceways may also be used. These wires must be continuous. They may not have splices between the utility connection and the meter con-

lowing types: RUW, RHW, USE, TW, THW, THWN, XHHW, and lead-covered. The local inspector might require additional protection.

METER AND SERVICE DISCONNECT

Usually only one service drop or service lateral is permitted per building. There are exceptions to this rule, but they do not apply to single dwellings. Figure 4-19 shows a mock-up of a mast, meter base, and conduit. The service disconnect would be on the first floor in cottages, commercial buildings, and dwellings with no basement. In this case, the cable might have to be brought out of the meter base and turned 180° and carried through the building wall to the service disconnect. See also Fig. 4-20.

In other cases, the disconnect is on the outside adjacent to the meter. It can also be combined in one cabinet with the meter base. Where there is a roof overhang all around the building, the service mast and head must extend through the overhang at a location to allow the conduit to lie flush against the building exterior wall. The mast must be high enough to provide the 10-foot minimum clearance. If the mast is quite tall to provide this clearance, a guy wire must be installed to anchor the mast. This arrangement is needed to take the strain of high winds, snow, and ice on the service drop. The conduit must be at least 1¼-inch trade size. It must also be anchored securely to the building structure below the overhang. Commercial buildings have specially formed galvanized angle iron masts with anchor bolts through the concrete block wall with 4-inch square plates on the inside under nuts and washers.

Conduit looks best, but cable can be used below the meter through the wall and into the service-entrance panel. You can also use service-entrance cable above the meter from the service head to the meter base, provided there is enough height to maintain the 10-foot minimum of the service drop. The choice is up to you (with the approval of the inspector). When a service mast must extend through the roof overhang, this mast must be conduit with a service head included. This is to provide the necessary 10-foot clearance of the service drop.

The service disconnect must have enough capacity to handle the current draw of the system. If the system is designed for 100-amp, 125-amp, or 150-amp capacity, then the disconnect or circuit breaker must have the same capacity. In addition, the wires or cable serving the system must have like ''ampacity.'' Refer to the tables in the appendix giving the ampacities of wires and cables. Some load centers incorporate a main breaker and breakers for

Fig. 4-21. Close-up of breaker panel. Upper arrow shows a double breaker for furnace/air conditioner. Lower arrow shows the neutral bar. The small screws accept the neutrals from the circuits. The medium screw is for the range circuit and the large screw is the service neutral connection.

individual circuits. In areas where permitted, this assembly can be installed outside. Weatherproof panels are manufactured so individual circuit breaker panels can also be installed outside.

The Service-Entrance Circuit Breaker (Main Circuit Breaker)

Overcurrent devices must be installed to disconnect the utility lines from the feeders for working on equipment beyond that point (the distribution panel, etc.). The other reason for the overcurrent device is to protect the wires between the main breaker and the distribution panel. These wires must be protected from currents in excess of their ampacity that is determined by their size and type of insulation. The wires from the service head all the way to the distribution panel are thus protected from overcurrents.

The common form of overcurrent protection now used is the circuit breaker. See Fig. 4-21. Breakers have the advantage of being able to be reset instead of being replaced. Breakers are available in 15-amp to 50-amp sizes (in 5-amp increments) and from 50-amp to 100-amp (in 10-amp increments). Thereafter the ratings are 110, 125, 150, 175 and 200 amp. Service-entrance equipment is available in various capacities to meet the needs of the installation. For dwellings, the range is from 50 amp to 225 amp. They are also available with space for a range from two breakers to 42 breakers.

If the service-entrance equipment and the main breaker or fuses is at some distance from the distribution panel, the wires between these two panels are called feeders. This might be where the main service breaker or fuses are outside in the cabinet with electric meter. The main overcurrent devices will protect these wires. *Note:* The larger sizes of wire used in

these cases must be connected to the equipment using pressure-type connectors. These connectors are those with an opening into which the bared wire end is inserted and a setscrew is tightened down on the wire, thus making a very secure connection. Soldering is prohibited.

When wiring houses, the neutral wire is grounded. See Figs. 4-22 and 4-23. The ground is usually a metal cold-water pipe from a city water supply. This is a very good ground. If you use service-entrance cable for your installation, the neutral will be bare. The insulated wires are in the center and the bare neutral made of fine

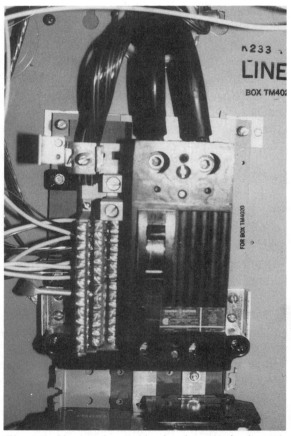

Fig. 4-22. Close-up of neutral bar in a large 42 circuit panel.

already taken care of by the manufacturer. The meter base is supplied by the utility and installed by the contractor or yourself. This is part of the service assembly you can install.

The Fused Service Entrance

The other type of overcurrent device is the fuse used in older installations. See Fig. 4-24. This older type of service-entrance panel consisted of a 60-amp enclosure and one or more "pull-out" fuse blocks (as in Fig. 4-25). Larger-capacity panels had more pull-out blocks to accommodate an air conditioner or other heavy amperage appliances such as an electric water heater or clothes dryer. The main fuse block would have two 60-amp cartridge fuses that protect the lighting circuits that have plug fuses. The fuse blocks for the range and other large appliances would be directly supplied from the main bus bars and would not be fed through the main fuses.

This type of service-entrance panel is available in sizes up to 125-amp capacity. Figure 1-4 in Chapter 1 illustrates a service-entrance panel of 125-amp capacity. Shown are pull-outs for the main, the range, and the air conditioner. Shown in use are eight circuits (four 15-amp and four 20-amp). The 15-amp "fustats" have hexagon windows to distinguish them as being 15-amp or less capacity. There are two spare circuits that are available for additions.

The disadvantages of the fuse-type panel is that spare fuses (or fustats) must always be on hand (both 15-amp and 20-amp sizes) to prevent the possibility of having to shop for fuses at such times as midnight, Sundays, or holidays. The fuses in the photo are the slow lag type having a special thread on the base (type S fuse).

Fuses are hard to change and require caution when you do so. The cartridge fuses on the

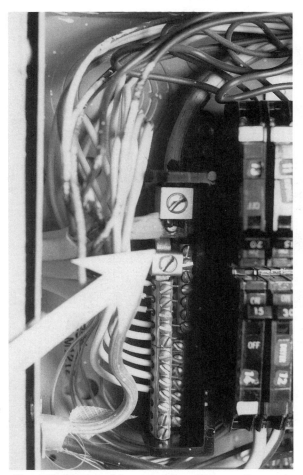

Fig. 4-23. Close-up of the neutral bar with connections made to it. The large screw accepts the neutral from the service entrance cable.

wires is wrapped around the two insulated wires. When a connection is made at either end, this neutral is unwound from the other two wires and formed into a large wire by twisting the strands tightly together. This and the other two are then inserted into their respective pressure-type connectors. The neutral wire, because it carries less amperage than the two hot wires, can be one size smaller, provided they are larger than No. 6. In service-entrance cable, this is

Fig. 4-24. Fused service-entrance panel, 125A, using Fustats instead of standard fuses. Upper right is the main fuse block (60A). Lower right are air conditioning and dryer fuse blocks (30A each). Eight circuits are fused, plus two spare fuse sockets.

pull-out blocks are expensive. The 30-amp fuses cost about $.60 each and the 60-amp main fuses cost about $1.25 each. In all other respects, fused service-entrance panels and distribution panels are most satisfactory.

I must make it very clear that any wiring that is required to be protected by an overcurrent device *must* be protected by the correct size of fuse or circuit breaker. And vice versa, a circuit breaker or fuse of a designated rating must only have wiring of the correct size and insulation material connected to it. Example: No. 14 wire may only be protected by a fuse or circuit breaker of 15-amp rating or less, not by a 20A fuse or breaker. For more complete information refer to the 1987 NEC Article 230, Services, and Article 240.

Overcurrent Protection

When deciding on the size of the service-entrance equipment, allow for expansion of needs as the family grows and the purchases of additional appliances. Larger entrance and distribution equipment costs more, but it will cost less than removing and replacing the small ca-

Fig. 4-25. The MAIN *pull-out fuse block. Holds two 60A cartridge fuses.*

pacity equipment with one of proper size. Also, the complete installation would have to be de-energized to make the changeover.

INSTALLATION AND GROUNDING OF EQUIPMENT

All wiring systems must be grounded for the protection of persons coming in contact with any metal parts carrying current. Grounding is also necessary to protect the wiring system itself from faults such as defective workmanship or materials, defects developing due to wear or age, and in case lightning strikes utility lines anywhere near the building, causing wiring to melt and switches and outlets to short.

Items To Be Grounded

Referring to Fig. 1-5 and Fig. 2-2, note that it is especially important to not only ground the metal parts not carrying current but the neutral (grounded wire) at the service-entrance panel where it is connected to the neutral bar. If the service entrance is combined with the distribution panel, the individual circuit neutral wires will also be connected to one neutral bar (a multiple connection point).

Methods Of Grounding

Referring again to Fig. 1-5, notice that the grounding conductor must be carried from the service-entrance panel to a good earth ground. In the case of Fig. 2-2, the city underground metal water supply pipe is an excellent ground. Be sure to attach the ground clamp on the street side of the water meter, otherwise, a jumper wire with a ground clamp on the water pipe on each side of the water meter must be installed. This is necessary to maintain a proper ground for the system if the water meter were to be removed for replacement or repair by the city. If there was no ground jumper wire with the water meter removed, the electrical system would be without a ground. This is exceptionally dangerous.

If no water pipe is suitable or available, a metal gas pipe could be used, but only with the approval of your local inspector *and* the gas utility. If you cannot use the gas pipe, you could use a well casing (metal) but not a drop pipe in a dug well. (The well might go dry and will not provide a ground in this case.) Any underground pipe of metal not having a corrosion-resistant coating can be used if more than 10 feet are underground and in direct contact with the earth. The metal building framework of a building may also be used.

Made Electrodes

If no suitable ground is available, you must use a *made electrode*. Approved ground electrodes are:

▶ A rod driven 8 feet into the ground. If made of steel, it must be ⅝ of an inch diameter minimum and be copper plated. If made of nonferrous metal, it must be at least ½ inch in diameter.

▶ Pipe or rigid conduit of ¾-inch trade size, either galvanized or metal coated, driven 8 feet into the ground. If the rod or pipe cannot be driven 8 feet into the ground, drive it in at a 45° angle. In all cases, the top of the electrode must be below or flush with ground level.

▶ Plate electrodes having two square feet of surface exposed. Steel plates must be ¼ of an inch thick and nonferrous plates 0.06-inches thick.

▶ Caution: lightning rod grounds should *not* be

used as system grounds. The plate-type electrode should be buried at least 18 inches below the surface.

The Grounding Wire

The grounding electrode conductor (*grounding wire*) should have no splices. It should be continuous for its entire length from the service-entrance panel neutral bar to the ground electrode (water pipe, rod, plate, etc.). This grounding wire can be of copper, aluminum, or copper clad aluminum. It must be at least No. 6, if stapled to the building. It can be insulated, bare, stranded, or solid. Smaller sized grounding wire can be used, but is has to be in conduit and the requirements are stringent. It is easier and less expensive to use a minimum of No. 6 or No. 4 and staple it. Both the entrance enclosure and the ground clamp must have pressure-type connectors for the grounding wire connections.

Grounding Devices

There are various types of grounding methods because of the design of various devices requiring grounding. They are as follows: a clip to push onto the edge of a wall box, a machine screw inside the wall box, the ground clamp setscrew, and a green hex head screw on devices such as switches and outlets. Although not thought of as grounding devices, locknuts, grounding bushings, and grounding wedge lugs really are for grounding. Another grounding device is the cord plug with the longer U-shaped third blade. This longer third blade makes a ground connection before the other two current-carrying blades make contact when inserted into the outlet. It is also the last to break contact when removed. *Never cut off the third blade.*

Some outlets and switches have a spring-type mounting strap-screw assembly that eliminates the need for a ground wire to the green hex screw on the device. This is used where the box is recessed in the wall and contact with the box is only by means of the mounting screw going through, say, an outlet mounting bracket. This is a very loose contact and is not good enough. The device has plaster ears that maintain its position so that it is flush with the wall surface. That is why the outlet mounting bracket will not be against the metal wall box surface.

If you try to move such a device sideways, it will be obvious that there is poor contact. This is why the strap-screw assembly is required. Otherwise, you must use the wire from the hex screw to the box-grounding-screw method. For obvious reasons, nonmetallic boxes cannot be grounded. Be sure to ground all devices. Any grounding wire must be the same size as the current-carrying wires. Grounding of all equipment *must* be continuous from the last metal enclosure all the way back to the service-entrance enclosure.

No grounding wire can be disconnected by a switch unless all wires carrying current are *simultaneously* disconnected from the devices themselves. The metal boxes must be tied in by means of the special green hex-head grounding screw that goes in the special tapped hole in the back of the box. Make sure the grounding is continuous back to the entrance panel.

UTILITY RULES AND RESTRICTIONS

Each individual utility has its own rules and will enforce them by withholding service to the property until you comply with the rules. For instance, no service will be supplied until a valid permit has been obtained by you from the governing body concerned. There are rules that govern overhead service drops and underground service laterals. The location of service equipment on the outside of the building will be determined by the nearest connection to the utility's lines. Any additional poles needed will be charged to you and they are expensive. Sometimes underground service laterals can be routed around obstacles. Underground wiring might be charged to the homeowner.

Some areas have passed laws prohibiting overhead wires and require that all utilities be underground. If you now have overhead service, it will be allowed to stay if you are only upgrading your wiring. Check with the utility before relocating the meter, mast, and service equipment any distance from the original location.

ENTRANCE OF CONDUCTORS TO THE BUILDING

Underground conductors come out of the ground either through a long sweep conduit bend or might just enter the vertical conduit directly into the meter base. From there, they might go down and through the basement wall to the service-entrance panel there. Overhead conductors come down to the meter through the service head and conduit to the meter base, and from there also down to the basement. In a dwelling with no basement, the service enters the building near the meter base to a utility room or other room on the first floor to reach the service equipment.

The Service Head

Conduit or cable must have protection from the elements so that nothing gets into the conduit or cable (especially moisture). The service head is designed to do just that. The head is a curved cast fitting; one is used for conduit and another is used for cable. There is still another for Thinwall tubing. It is in a gooseneck shape, and the wire opening faces down at an angle to prevent moisture from entering. Each wire has a separate hole in the plastic insert. Although cable can be bent into a gooseneck (provided it is taped and painted and used without a service head), a better looking installation results and the top of the cable is anchored. Add one, because they are very reasonable.

Drip Loops

Generally, the attachment point of the service drop is below the service head. Some installations have the drop *above* the head. This might cause problems with water droplets finding their way into the cable or conduit. To overcome this condition, the service head wires must drop down lower than the service head by 6 to 8 inches and then make a U-turn back up to the attachment point of the utility wires to the building. This lower loop allows water to drip off at bottom of the loop and not enter the service head or cable.

MOVING THE ELECTRIC METER OUTSIDE

Moving an electric meter outside might seem difficult to do, but actually it's quite simple. The time to do this work is when either updating or completely rewiring your house. Refer to Chapter 13 for detailed instructions. If you can rough it for a day or so, borrow electricity from your neighbor using a heavy-duty electric cord, or perhaps you could arrange for a temporary power connection.

The actual changing of the service-entrance panel and accessories can be done without disrupting power until the actual changeover connections are made. At this point, make arrangements with the utility as to when you want the new service connected. In some very old houses, the service wires entered the second floor at ceiling level and the meter might be in a second floor hallway.

Assemble the service mast/meter base/conduit/LB and make the necessary hole through the foundation wall for the nipple that will go through the wall. Now get help to mount the assembly, using two hole pipe straps. The new service-entrance/meter base assembly can be mounted close to the old conduit. *Caution:* When mounting the new assembly, avoid the service drop wires. It would be best to tape any bare spots on the service drop wires before mounting the new assembly.

The old conduit is shown as ¾-inch. It has a gooseneck bend at the entrance through the wall into the basement. Special fittings to make a 90° turn were not available. There are now Condulet fittings that allow the conduit to lie flat against the building wall. If you are using an LB in the new assembly, the changeover should take no more than one day unless you run into problems. Consider all conditions you might run into and try to solve them before starting the job.

If you continue below the meter with conduit, use an LB angle conduit fitting to make the turn to go through the wall into the basement. This whole assembly can be made up in advance and installed in one piece. You will really need help for this assembly.

Insert the nipple that is screwed into the LB through the house wall holding the assembly horizontally. Now rotate the assembly to vertical and anchor with suitable pipe straps. Be careful to avoid contacting the service drop wires. They are still hot. Have help.

Before mounting the assembly, you can install the wires from the service head down to the meter base. A 3-foot length of wire is needed, extending from the service head for the utility to connect to the service drop. Tie these ends to the mast to prevent accidental contact with hot wires. Connect the other ends to the meter base terminals: red wire to left terminal, white wire to neutral terminal, and black wire to right terminal. Figure 4-20 shows the meter base and wiring connections. Wires from the service drop to the meter base and from there down and inside to the entrance panel may be of aluminum.

Refer to the NEC Table 310-16, Note 3. It states: Three-Wire, Single-Phase Dwelling Services, 100A ampacity for No. 2 aluminum wire and 100A ampacity for No. 4 copper wire. This is for either cable or individual conductors at 90 degrees Celsius (194 degrees Fahrenheit) temperature rating. If you are using cable below the meter base and into the basement or utility room, you should mount the service head, conduit and meter base before continuing on into the house.

If you use conduit all the way, assemble everything before installing. The wires can be pulled in beyond the meter base later.

If you have an installation similar to the preceding situation, proceed as follows. Fully position a *wooden* ladder so that you can reach the connection of the entrance wires to the service drop wires (three or two depending on the old installation). *Do not use an aluminum ladder.* Using side-cutting pliers with approved insulation on the handles, cut each wire separately between the service head and the service drop 6 inches from the clamp connector on the service drop wires (Figure 13-34 explains how to do this procedure). You will not receive a shock if you are careful to cut only *one* wire at a time.

Touch nothing but the wooden ladder (dry conditions) and the insulated plier handles. Cut each wire in turn and bend them down out of the way of the service drop as you cut them loose. Do not touch the cut ends of the wires until all are disconnected from the service drop. Do *not* cut the wires from the pole to the house.

When all the wires have been disconnected, you can remove the conduit. It is best to cut the conduit at the gooseneck. *Caution:* look up before you start cutting to see if everything has been disconnected at the top. Now cut right through the conduit and wires. Use care in removing the assembly so that you do not hit the cutoff wires on the service drop—they are hot.

It is best to have help when removing the assembly so as to avoid hitting the hot wires. The assembly might be heavier than you think. Everything inside can now be removed. Remove the meter so that you can turn it over to the utility when they come to make the changeover. The meter department will set a new meter and the line crew will connect the new service-entrance equipment. If the meter was inside, all the wiring will be very old and outdated. In this case, you will be installing a new service-entrance disconnect/circuit breaker assembly. This can include the distribution panel with its breakers for the individual branch circuits needed. Because the new panel location might be different, you might need more Condulets and conduit nipples to continue the conduit into the panel.

If all the old wiring is to be scrapped, remove the old fuse panel and separate disconnect, if there is one, so that you can mount the new service in place. When this equipment is installed and wired, and the utility has made the connection to its lines at the service mast, you might need to call for an inspection from the municipality. This depends on the inspection department. Check with the inspector.

If the entrance switch/breaker or fuse panel is separate from the distribution panel, connect them with a 2-inch nipple, locknuts, and insulating bushings. The more free area in the conduit, the easier it is to pull in the wires. It might be easier to use service-entrance cable if the distribution panel is some distance away from the service disconnect. Cable works easier if there are many turns.

The service disconnect must be directly inside the building. The NEC does not allow long runs of entrance cable inside a building without being protected by a fuse or breaker. This is very important. If it is necessary to mount the service disconnect some distance from the meter location, run the entrance cable on the outside of the building; then go through the outside wall directly into the service disconnect. These back and forth references to conduit and cable are to show that both are accepted by the NEC. Check with the local inspector in advance. Conduit is always approved because there is no better raceway.

If you have inspected the present wiring and feel it is in safe operating condition, you might need only replace the service-entrance/distribution panel assembly. All knob-and-tube wiring must be replaced because it is very old. This type of wiring has rubber insulation covered with cotton fabric. In hot areas, the rubber deteriorates, cracks, and falls off in pieces. Remove all of this wiring.

PROBLEMS WITH OLDER HOUSES

Older houses have knob-and-tube wiring and perhaps the meter is upstairs. Prepare for a lot of hard, dirty work. There will probably

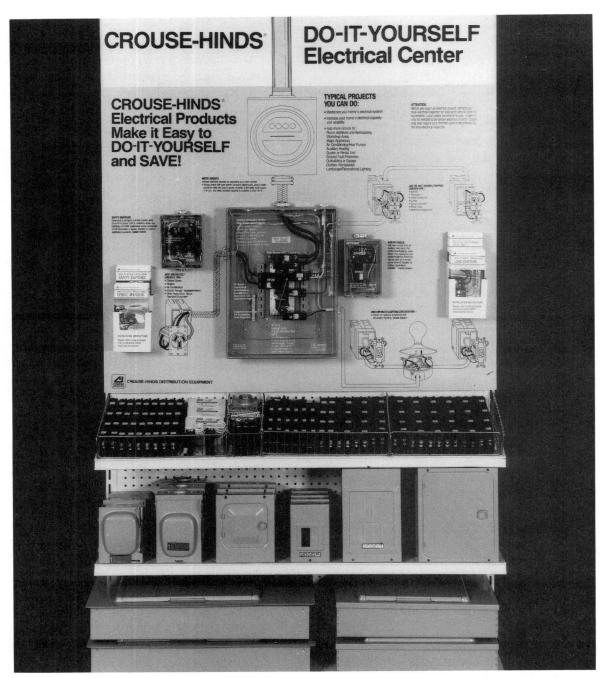

Fig. 4-26. Shows a "Do-It-Yourself" electrical center. Provides all wiring materials needed to install service-entrance equipment. Available at home centers. (Courtesy Crouse-Hinds)

be two to six circuits. Much depends on the size of the house and when it was wired. Many fuses in older houses were in openfaced fuseholders mounted on the basement ceiling or wall or on the second floor.

Old knob-and-tube wiring (open wiring on porcelain insulators) in the attic can be used if the insulation is in good condition, the wiring is strung tightly so that it does not sag and touch any flammable material, and if it is in use and is fused at 15A, no more. All new wiring should be Romex or BX to comply with local codes.

Review your plan again and complete it to give you a more workable guide to follow when you start the actual work. You will need a plan for the first floor and basement. If there is a second floor, you need an additional plan. See Fig. 4-26. Chapter 12 takes you through a complete rewiring on an existing building.

Chapter 5

The Fuse/Circuit Breaker Panel

The *distribution panel*, also called the fuse panel or the circuit breaker panel, receives its current from the service-entrance panel. These two panels are often combined in one cabinet or enclosure. In single-family dwellings this is usually the case. The main breaker is at the top of the panel with the individual breakers taking up the rest of the cabinet. A fused distribution panel has a main fuse pull-out block and one or two other fuse blocks for a range, water heater, or dryer. There will be eight to twelve individual fuses for the various circuits supplied by them.

THE ENCLOSURE

The service-entrance panel with its circuit breaker or fuses can be mounted outside or inside the building. Many situations arise where the distribution panel cannot be adjacent the entrance panel. In this case, the service-entrance cable should be used to connect between the two panels. This cable must have the same ampacity as the cable feeding the entrance panel from the meter. It must be four-wire cable having one black, one red (or black with red stripe), one white, and one ground. This last ground is needed to provide grounding for the metal cabinets and other noncurrent-carrying parts.

Branch circuits are classified by type:

▶ lighting circuits
▶ appliance circuits
▶ heavy appliance circuits (range or dryer)
▶ laundry circuits

The number of outlets on lighting and appliance circuits is not limited by the NEC; use six to ten for each circuit. After you have decided on the number of circuits that you want (both single- and two-pole), purchase a panel with space for 10 to 25 percent extra. Later, an additional panel can be added beside the original. Figures 5-1, 5-2, and 5-3 illustrate panels and service equipment.

OVERCURRENT AND SHORT CIRCUIT PROTECTION

Modern residential installations tend to use circuit breakers instead of fuses. There is nothing wrong with using fuses, but you must have spares on hand for when you need them. Breakers, once not ideal, have been improved so that they are now very reliable.

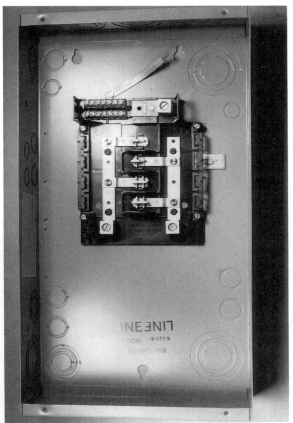

Fig. 5-2. Circuit breaker panel with one, 1-inch breaker installed. This is a mock-up to show the appearance without all the breakers and wiring installed. The neutral bar is at the top.

Fuses

Fuses provide overcurrent protection in addition to short-circuit protection. See Figs. 5-4 through 5-6. The amperage (current flow) the wire can handle is limited by the wire size and insulation type. Refer to NEC Table 310-16 to 310-19, Note 3, which lists Three-Wire, Single-Phase Dwelling Services, Conductor Sizes, and Types for both aluminum and copper wire. Wire or cable has the designation printed or impressed on its surface at regular intervals, such as "14-2 w/grd." or "12-3 w/grd." along its

Fig. 5-1. Distribution panel to hold twelve 1-inch-wide circuit breakers. The main can be installed in this panel also.

Fig. 5-3. Illustration of another service entrance panel with breakers installed. Another breaker is shown at top.

Fig. 5-4. Pull-out fuse block, complete with two 60A fuses.

full length. The safe operating voltage, also determined by the insulation, is printed or embossed along with the wire size and number. This designation might be "300V" or "600V."

Plug Fuses

The standard plug fuse has an Edison-base thread just like a regular light bulb. This fuse comes in the following amp rating sizes: 10, 15, 20, 25 and 30. It consists of a hollow porcelain or glass body with brass threads (like a light bulb) on the outside. On the top is a window

Fig. 5-5. Shows a fuse adapted to fit 60A fuse clips using reducers. Fuse can be 30A or smaller and still fit.

Fig. 5-6. Shown is a Fustat and adapter used in standard fuse panels for better overcurrent protection.

of glass or mica so you can see the *fusible link* inside (see Fig. 5-7). On the bottom is a contact that touches the bottom screw in the fuse panel socket. The fusible link inside connects the bottom outside contact with the screw shell. When the fuse is screwed in place, the circuit is energized. Fuses up to and including 15-amp capacity have a hexagon-shaped window. Larger-capacity fuses have a round window. Plug fuses of special low amperage are used to protect motors and are usually mounted directly on the motor itself. Other motors have an overload device built in.

Time-Delay Fuses

Electric motors, when starting, draw from three to six times their normal running amperage when at normal speed. If a motor draws 6 amps when running, it might draw 20 amps or more when starting. This will blow a 15-amp fuse. For just these situations the time delay fuse was developed. This type of fuse carries a 200 percent overload for 12 seconds and a lesser overload for 30 seconds. However, they blow immediately on short circuits.

Type S Nontamperable Fuses

To prevent the substitution of oversize fuses (defeats the purpose for which fuses are

designed), a nontamperable fuse was developed. This fuse is different from the standard fuse. The threaded part is smaller and the threads are finer. This prevents a coin from being put behind the fuse and also prevents bridging by means of foil of any kind. In addition, only 15-amp and smaller fuses will fit in a 15-amp adapter that is part of this fuse system.

Type S fuses are sold and manufactured by various companies, but they all have the same features. They have the time lag feature in addition to being nontamperable. To fit in the

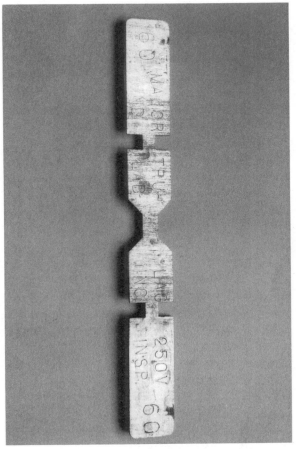

Fig. 5-7. A fuse link for use in a renewable cartridge fuse case. This link saves money over replacement of the complete fuse.

standard Edison base fuse holder, they are provided with an adapter that is screwed onto the fuse base. Then this assembly is screwed into the fuse holder. The fuse can now be removed for replacement, but the adapter cannot be removed because it has a spur sticking out of the Edison threads that digs into the internal threads of the fuse holder—thereby preventing removal. The adapters match the fuses as to ampere capacity (through 15-amp for one size, 20-amp through 30 amp for another size). Be sure you insert the correct adapter into the fuseholder because you have only one chance.

Cartridge Fuses

For conditions requiring fuses or time-delay fuses larger than 30 amps, *cartridge* fuses are used. This fuse is a cylinder with brass end caps and a fiber center. It somewhat resembles a rifle cartridge, hence the name. Inside is a fusible link and a dry powder to quench the arc when the fuse blows. To prevent overfusing, cartridge fuses are made in different sizes. Up through 30 amps, they are 2 inches long by $9/16$ of an inch in diameter. From 35 through 60 amps, they are 3 inches long by $13/16$ of an inch in diameter.

All these are rated at 250 volts or less capacity, meaning they can be used on voltages up to 250 volts. Large amperage fuses have knife-blade ends that fit between spring-loaded contact plates. Larger amperage and voltage fuses have larger dimensions mainly to contain the arc that forms upon blowing. Normally, you would buy a circuit breaker for your "main" disconnect (rather than fuses). Do not be concerned with the larger sized fuses. Service-entrance panels are available that have fuses as overcurrent protection for the mains; these are approved by the NEC.

CIRCUIT BREAKERS

Because of the circuit breakers' reliability and permanent feature, they have become the preferred type of overcurrent devices used today. Circuit breakers are marked with a setting which is calibrated at the factory and cannot be changed in the field. This setting represents the rating of a standard or time-delay fuse. See Figs. 5-8 and 5-9.

Breakers are selected by their rating or the tripping value. Fuses are selected by rating only. Because the results are identical—protecting the wires in the circuit—either type of overcurrent device is very satisfactory.

PANEL REPLACEMENT IN AN OLDER HOME

The replacement of a distribution panel in

Fig. 5-8. Two-circuit breaker panel with one double breaker installed. Will hold two 1-inch or four ½-inch breakers.

Fig. 5-9. Panel in 5-4 shown open without breakers installed.

an older home must depend on the capacity of the existing service-entrance equipment. If the service-entrance panel does not have sufficient capacity, both panels must be replaced. In this case, buy a combination service-entrance/distribution panel. Using a combination panel depends on the location of the meter in relation to the distribution panel inside. It is highly recommended that everything except the meter be located inside. Be sure to have the main breaker as close as possible to the point where the entrance cable enters.

Look the situation over and decide if the

point where the service enters the building will have room inside (basement or other selected location) for the combination service-entrance/distribution panel. If the location of the old fuse panel has ample room for the new combination panel, then install it there. The circuit leads from the old panel might have to be spliced or even go around to the other side of the new panel. Sometimes you need a junction box (about 8 inches by 10 inches) with a cover to provide a neat workmanlike installation for extensions of these circuit leads. If your work is neat and shows care in the installation, the inspector will be favorably impressed. This will make a big difference in the inspector's attitude.

Start by removing the old panel after first disconnecting the circuit leads. Tie up the leads out of the way to make room to work. Hold the new panel up in place. It is usually empty except for the main breaker or fuse blocks. The circuit breakers are not supplied and neither are fuses in a fuse panel. Breakers or fuses are selected according to needs. Don't forget to install ground fault circuit interrupters (GFCI's) for all bathroom, outdoor, sauna, pool, garage, and other hazardous locations. See Fig. 5-14.

If the enclosure is a combination, be sure to allow room for the entrance cable to come into the enclosure back or side. Enclosures have knockouts in the back so that conduit or cable can be brought through the outside wall directly into the enclosure. Basement enclosure installations would generally have the cable or conduit brought in through the top of the enclosure because of the closeness to the top of the basement wall. The enclosure should be at eye level for inspection and servicing; don't mount it too high. A first-floor location can use a flush enclosure, which is designed to be mounted between studs.

If the basement wall is concrete, drill holes

for anchors. The anchor size needed is ¼-20 A&J's. This means a lead anchor having a steel insert with a tapped hole in it having 20 threads per inch and accommodating a ¼-inch bolt. This anchor needs a ½-inch hole in the concrete for installation. Use a carbide drill bit and drill motor. Make the hole slightly deeper than the anchor. A star drill and hammer will work, but the carbide drill does not cost much and it saves time. Be sure to use goggles whenever you drill. Hold the enclosure level and mark for the anchor holes. I have found a china marker pencil best for such marking. The one I use has a crayon point and advances the point just like a mechanical pencil. This pencil (made by Listo) is available in any stationery store for about $1. Refills are also available.

When drilling with a carbide drill bit, be sure to guard against the bit wandering from the spot marked. Before starting to drill, mark over the original mark a vertical cross with china marker. That way you still have a remnant of the original vertical cross even though the hole has removed the original mark at the center. If the drill does wander, try to angle the drill bit and motor to try to move the hole in the right direction. If the wall is hollow, use ¼-20 toggle bolts. These are 3-inch to 4-inch bolts with a nut that has wings on it. They are squeezed together and inserted through the hole into the hollow part of the wall. The wings are spring loaded and open out after they are in the hollow part of the wall.

Because the hole size needed to insert the wings is larger than the bolt shank, there is room for some adjustment if the hole is off its mark. Remember that the toggle bolt must be put through the hole in the metal enclosure, then put through the hole in the wall and tightened. This is different than the A & J anchor, as the anchor is fixed permanently in the wall before

the bolt is inserted. Therefore, with the toggle bolt you have only one chance. If you insert the toggle bolt and for some reason need to remove it, you will lose the toggle part inside the wall.

Make all your work neat, plumb, and level. Everything will look better, work better, and the inspector will be pleased with your installation.

ACCESS TO THE PANEL

Easy, clear access to the electrical service equipment and distribution panel must be provided in case of emergency, for routine inspection, and for servicing. NEC Section 110-16 requires this accessability. While this section also requires lighting and head room for a large amperage installation, providing this for your residential installation makes good sense. If you have relocated the entrance of the cable or conduit into the dwelling, pick a suitable location for the equipment to be mounted inside. The service-entrance/main breaker panel should be installed immediately after the cable or conduit enters the building, because there is no overcurrent protection for the wires between the meter and the main circuit breaker. If necessary, the distribution panel can be some distance from the entrance panel. Refer to the section in Chapter 4 titled "Moving The Meter to The Outside."

CAPACITY AND CALCULATIONS

In this section, I have made capacity calculations for a dwelling built in 1940 in a suburb of Detroit. The building is 28 feet wide and 26 feet deep, front to back. It is a one-story frame construction with a full basement. (Refer to Figs. 12-1 through 12-4.) This example starts with the wiring installed when the house was built. Note that there were few outlets and the service was 115 volts and possibly 60 amps. To completely update this dwelling, all the wiring will have to be replaced. At the time of construction, the outlets, fixtures, and switches were not required to be grounded. The 1987 NEC requires that a minimum of 100-amp service be provided. Older houses with 60-amp service might be able to use this service equipment, provided the ampacity is adequate for the calculated load.

To start, assume—for making the calculations—that this is a new house under construction and the wiring has not yet been installed; also assume that power will not be needed for some time and that the house is not occupied. If when you actually rewire your own house, you might have to work on the wiring and live in the house at the same time. This can be done, but with some inconvenience. The work to be done first—in this case—is to change the service-entrance/breaker panel assembly and move the meter outside (if it is still inside). After that, you can rewire the rest at your convenience. In the above example, if you disconnect all power and need temporary power, either buy power from your neighbor or have a temporary service installed.

A note of warning—if you get power from your neighbor, even if you are good friends or especially if you are just new in the neighborhood, be sure to insist on paying. I suggest $1.00 per day. I have seen many cases of hard feelings over this situation.

ALTERNATE CALCULATION

Perhaps your dwelling has a floor area of 728 square feet (exclusive of unoccupied basement), no attic, and a small front stoop. Perhaps it has a 12 kW range, a 2.5 kW water heater, a 1.2 kW dishwasher, a 5 kW clothes dryer, and a 4.5 kW central air conditioner.

While I have included an electric water heater and an electric clothes dryer in the computation, I highly recommend that these two appliances be gas fired. Electric resistance heating (red hot glowing coils) that also includes electric space heating, is the most expensive and also the most wasteful of energy. While this book is about electricity, I still advise buying gas appliances. If the above appliances are already installed, use them, but when replacements are needed, buy gas-fired types.

The dwelling represented by the calculations in Table 5-1 can be supplied by a 100-amp service. NEC code requires two small appliance circuits for kitchen and dining areas and one circuit for the laundry. These three circuits must be 20 amps because of the heavy loads from toasters, ovens, and other heavy wattage appliances used in these areas. As shown in Table 5-1, the calculated load of 18,074 watts must

Table 5-1. Computed Load

	Watts
General lighting, 728 square feet @ 3 watts per foot	2184
Two 20-amp appliance circuits @ 1500 watts each	3000
One 20-amp laundry circuit	1500
Range circuit (at nameplate rating)	12000
Water heater	2500
Clothes dryer	4500
Central air conditioning	4500
Total	30184
First 10,000 watts @ 100%	10000
Remainder @ 40% (20,184 watts × 40%)	8074
Total	18074
Calculated Load: 18,074 ÷ 240 = 75.5 amps	

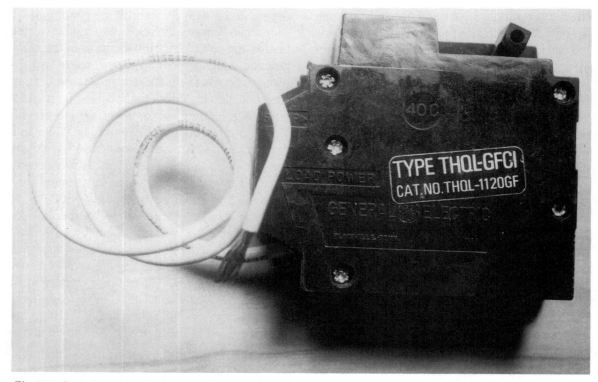

Fig. 5-10. Ground fault circuit interrupter (GFCI), 20A circuit breaker combination. Installs in place of a standard circuit breaker and protects entire circuit.

now be divided by 240V rather than 230V. This is required by the 1987 issue of the National Electrical Code.

GROUND FAULT CIRCUIT INTERRUPTER

Many times faults occur in wiring that are not great enough to blow a fuse or trip a breaker. While these faults are not high, they *can* cause damage. Individuals could be injured and death could result under some circumstances. These are generally small current leakages that give you more of a "tingle" rather than a greater shock.

Many fires are caused by these same small current leakages where heat (that results from the leak) is not readily noticeable. For these reasons, the ground fault circuit interrupter (GFCI) was developed. See Figs. 5-10 through 5-14. The GFCI interrupts a current of $5/1000$ of an ampere almost instantaneously. As you can see, this is a very sensitive and fast-acting device.

NEC Requirements

The NEC requires GFCIs in bathrooms, garages, and out of doors for pools, hot tubs, marinas, and any place where the danger of shock is very great. They are also required on new construction sites for connection of power tools, trouble lights and extension cords. Con-

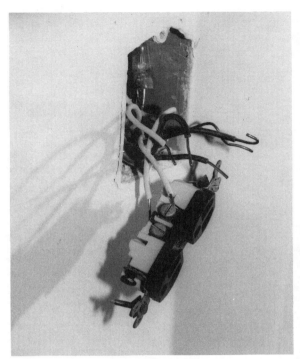

Fig. 5-11. *Installation of a GFCI. Old receptacle is being removed. All the wires will be checked out before new installation.*

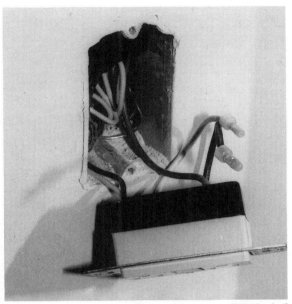

Fig. 5-12. *Leviton® ground fault circuit interrupter (GFCI) wired in circuit. It is not wired to protect the rest of the circuit.*

struction personnel are more frequently subject to shock because of defective equipment and wet or damp places where equipment must be used.

Types of GFCIs

GFCI devices come in three types: the plug-in receptacle, the circuit breaker, and the portable. The receptacle type is usually installed in the bathroom, a potentially dangerous place. This type will protect its own receptacle only or can be so wired that it will protect other receptacles on the same circuit if they are beyond that point. In other words, if the bath is the first receptacle on that circuit, it can protect any other receptacles *beyond* itself, away from the distribution panel. Be sure when in-

Fig. 5-13. *Completed GFCI installation. Device must be tested every two months. Push "test" button, then "reset" button.*

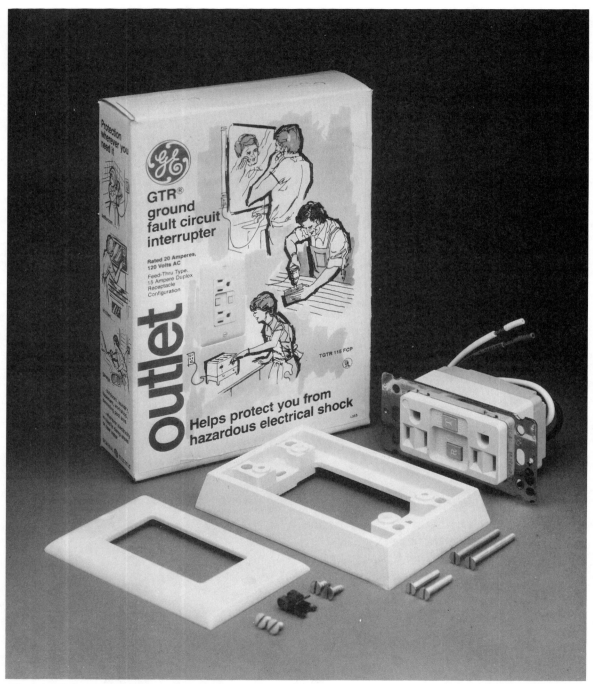

Fig. 5-14. High quality GFCI by GE. The raised adapter allows it to be installed in a shallow wall box. (Courtesy General Electric Co.).

stalling a GFCI to connect the leads on it marked "line" to the wires coming from the panel and the leads marked "load" to the wires leaving the receptacle box and feeding the other receptacles down the line. If the bath receptacle is second or third, it cannot protect "backwards" toward the distribution panel. When a "push-to-test" button is pushed, the reset button next to it will pop out. Push the reset button to restore service. Test the GFCI every two months.

The circuit breaker type is mounted in the breaker panel just as are standard breakers. It has an additional white wire for grounding its internal mechanism to the panel neutral bar. This type will protect all outlets on its circuit because it is first in line. This type of breaker must be reset at the panel. The breaker type is unsatisfactory for circuits having very long runs, because voltage drop can cause the breaker to trip. The portable type of breaker is intended primarily for construction sites. Portable power tools and extension cords can be plugged into this special panel and thus the user is protected. All types come rated 15A or 20A.

Chapter 6

Branch Circuits

Lighting and small-appliance circuits are used in homes for ceiling fixtures, wall fixtures, and wall receptacles (outlets). These circuits are rated at 15 or 20A at 120 volts. The 15A circuit can be wired with No. 14 copper wire, but the 20A circuit *must* be wired with No. 12 copper wire. The 15A circuit can also be wired with No. 12 wire if you prefer.

Do not use aluminum wire inside the dwelling except to feed the main breaker and distribution panel. The branch circuits are also defined as "150 volts or less to ground" by the NEC. This means that the voltage between the hot wire and the equipment ground is 150 volts or less; in this case it would be 120 volts.

Even though a dwelling having 240V service has no hot wire with a voltage more than 150 volts to ground, each wire has voltage to ground of 115 to 120 volts. This comes within the definition stated above.

CLASSIFICATION OF CIRCUITS

Circuits are classified by the NEC according to the rating of the overcurrent protection, provided as follows: 15A, 20A, 30A, 40A, 50A, and 60A. These are general circuits, not individual circuits. Even though the wire itself has a higher ampacity than the overcurrent device, the circuit still is designated by the rating of that device. An exceptionally long run from the distribution panel would require the next larger wire size to minimize the voltage drop and hold it to less than 5 percent. Refer to NEC Tables 310-16 through 310-19 for ampacities of various wire sizes and insulation types. The 15A and 20A circuits are commonly used in a dwelling as explained above. Circuits designated as 25, 30, 40, and 50 amps are designated as individual branch circuits for connection of large permanent equipment such as a range or clothes dryer.

Individual Branch Circuits

Such circuits are used for well pumps, sump pumps and garage door openers. Other individual branch circuits are used for electric water heaters, clothes dryers, central or window air conditioning, electric ranges, and furnaces. Each of these circuits can supply only one piece of equipment and must be wired and protected by an overcurrent device appropriate for the ampere draw of such equipment.

Depending on the nameplate rating of the appliance served, the circuit will be rated and protected accordingly. An electric clothes dryer rated at 4500 watts (4.5 kW) requires a 30A breaker and is wired with No. 10 wire. Electric water heater circuits must be protected at 125 percent of the nameplate rating, therefore, a water heater rated at 4500 W × 125% = 5625 W, divided by 240V = 23.4A. Hence, overcurrent protection of 30A is required.

Air-conditioning systems are marked with the "minimum circuit ampacity" and the maximum fuse size. As an example: minimum circuit ampacity = 21.2A; maximum fuse size = 30A. For electric ranges, NEC Table 220-19 gives for not over 12 kW rating the maximum demand of 8 kW. For ranges or cooktops and wall-mounted ovens over 12 kW through 27 kW

combined ratings, the maximum demand should be increased 5 percent for each kW or fraction thereof that exceeds 12 kW. Table 6-1 shows an example of the calculations.

Other individual branch circuits supplying appliances such as well pumps, sump pumps and garage door openers should be wired with No. 12 copper wire protected by a 20A breaker or fuse. Try buying a quantity of wire, the service-entrance/distribution panel, and the circuit breakers in one place for a possibly considerable discount. An order for 1000 feet of cable or BX (plus the other items) is different than buying only 25 feet of Romex. If you are buying service-entrance cable (because you will not need much over 25 feet), it is best to try a local hardware store or home improvement center. This cable runs up to $6.75 per foot. Try to purchase these items at an electrical wholesaler. They just might sell to you and the price could be even lower. Shop around and save money.

To avoid buying too much cable, I take a cord, clothesline or any other line and actually run this as if it were the cable. Start at the outside where the cable would begin—at the service head above the meter location—and run it down and into the basement or other service-

Table 6-1. Ampacity Calculations

Cooktop	8 kW	5% × 7 = 35%
One wall-mounted oven	6 kW	8 kW × 35% = 2.8 kW
One wall-mounted oven	5 kW	8 kW + 2.8 kW = 10.8 kW
Total	19 kW	

10,800 ÷ 230 = 47A

Use No. 8 copper wire

Fuse for 50A

entrance location exactly as if the cord were the cable itself. After marking the length needed, you will be able to estimate the amount of cable to buy. Be sure to add 10 to 15 percent for bending the cable and stripping the ends. The power company requires 3 feet at the service head for their tie-in. At the service-entrance panel, be sure to allow enough wire length to reach all terminals in the panel. Sometimes the neutral terminal is at the bottom of the panel and the neutral wire must reach down to that point. This means having excess length on the two hot wires. This occurs in many cases, but having a wire longer is always better than having a wire too short.

Color Code for Branch Circuits

The *grounded* conductor is always white or a natural gray. The equipment *grounding* conductor always has green colored insulation or continuous green color with one or more yellow stripes—or it is bare. The *ungrounded* conductors can be any color except the above colors that are reserved for their special uses. This is extremely important and must be made very clear. The hot wires can be any other color, such as black, red, blue, brown, etc.

Because larger sizes of wire do not come in white, it is necessary to paint or tape (with white paint or tape) the ends of the wire where they can be seen in panels and other places so that this wire is recognized as the neutral or grounded conductor. The NEC requires this method.

Receptacles and Cord Connectors

All new receptacles installed on 15A and 20A circuits must be of the grounding type. If a means of grounding does not exist, an ungrounded type can be installed. *Caution:* Never install a replacement grounding type receptacle on a circuit that is not grounded. This gives a false sense of safety and it is dangerous if a defective grounded appliance or tool is plugged in to that receptacle. All tools and appliances that have attachment plugs must never have the grounding blade removed. This defeats the purpose of the whole grounding system and is very

Fig. 6-1. An incandescent dimmer being installed in a wall box. The terminals will be wrapped with electrical tape.

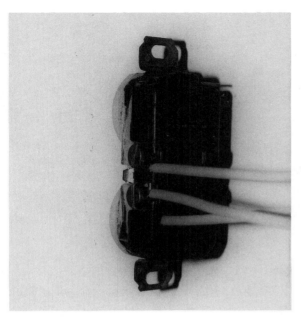

Fig. 6-2. Method of connecting three wires in a junction/wall box. Two wires are under terminal screws. Two more wires can be inserted in the push-in holes in the receptacle back, allowing four wires to be connected to this.

dangerous. In addition, the NEC prohibits this. See Figs. 6-1 and 6-2.

Ranges and clothes dryers can be wired with service-entrance cable, provided the cable starts at the service-entrance panel and is continuous all the way to the appliance receptacle. If the cable comes from a subpanel, it must be four-wire cable. One wire, the green or bare wire, is to ground the equipment. At each end, if the cable comes directly from the service-entrance panel, the neutral conductor (which is bare, stranded, and is wrapped around the two insulated wires) must be unwrapped from the other wires and twisted to make a conductor. For these two appliances only, the neutral can be used for both neutral and grounding purposes. This is the only exception to the rule of neutral and grounding wires being separate parts of the complete electrical system.

Wire Sizes and Ampacities

Wire sizes range from No. 18, used in lighting fixtures, to No. 3/0 and much larger for commercial and industrial installations. You can use wire sizes from No. 14 for individual circuits to, possibly, No. 3, which has ampacity of 110 amps for dwelling service. Any feeder wires or cable between the service-entrance enclosure and the distribution panel enclosure must be the same size as those of the service-entrance wires or cable from the service drop point of attachment. The NEC Table No. 310-16 lists the ampacities of various sizes of copper and aluminum wires by insulation. Note 3 to Tables 310-16 through 310-19 lists applicable ampacities for single dwelling units.

Types of Insulation

Code Table 310-13 lists conductor application and insulation types. XHHW insulation is used in making service-entrance cable. This is a very popular construction now. Because it is rated for wet and dry locations, it can be used outside for the cable run from the service drop, to the meter base, and from there into the building to the service-entrance panel. In addition, it can be used for the feeder to the distribution panel, for the electric range, and for the electric clothes dryer.

Types TW and THHN are used in making Romex and BX and also for insulating solid or stranded wire. Usually, the type available in the area is suitable and approved by the licensing authority.

ADEQUATE WIRING

A dwelling wired many years ago, although it might have been sufficient capacity then, is surely not adequate for today. Even with energy conservation measures, there still is a need to have adequate wiring. The minimum ampacity recommended for rewiring is 60A. For a dwelling with a floor plan of 1500 to 2500 square feet, 100 amps is not too large.

The extra cost of ample 125A service equipment, including the required heavier wiring, is nowhere near the expense and labor of removing almost new lower ampacity equipment and buying new larger equipment. In either case, the cost of branch circuit breakers will be the same (more might be needed) because 15A,

Fig. 6-4. Cut-away of standard circuit breaker.

20A, 30A, 40A, and 50A breakers will still be used.

The main breaker, the enclosures, and the service-entrance cable or conduit and wires will be larger. The enclosure will be larger to accommodate the larger main breaker and provide space for more branch breakers. See Figs. 6-3 and 6-4.

Wire Sizes in Relation to Distance

The maximum voltage drop now allowed in a circuit is 5 percent—a drop of more than 5 percent results in poor operation of such equipment as motors and electronic equipment. This can damage the equipment. If the voltage drop calculates at over 5 percent, use the next larger wire size. The basic formula for figuring the voltage drop in an alternating current circuit is shown in Table 6-2.

For example, VD = 4.57V; 4.57V divided by 230V = 0.0198% = 2% voltage drop means this voltage drop is satisfactory. This calculation comes in handy when figuring the size on farm wiring between buildings.

Fig. 6-3. Service-entrance panel for a large house. The main lines are quite large; panel will have 24 circuits.

Table 6-2. Voltage Drop Formula

$$VD = \frac{2 \times r \times L \times I}{\text{circular mills}}$$

VD = Voltage drop

r = Resistivity of conductor material (12 ohms per CM ft. for copper and 18 ohms per CM ft. for aluminum wire)

L = One-way length of circuit in feet

I = Current in conductor in amperes

Example: 230-volt, 2-wire heating circuit. Load is 50 amp. Circuit size is No. 6 AWG THHN insulation. Wire copper and the one-way circuit is 100 ft.

Substituting in the formula, Circular mills = 26240

$$VD = \frac{2 \times 12 \times 100 \times 50}{26,240 \text{ (Table 8, Ch. 9)}} \qquad VD = \frac{120,000}{26,240}$$

Noninterchangeable Outlets and Plugs

Receptacles that are on circuits having different voltages or special uses must have a configuration (arrangement of slots or blades on a plug) for accepting an attachment plug designed only for that voltage or special use. This prevents damage to appliances or other electrical equipment plugged in accidentally. I once plugged a percolator into a 230-volt outlet that had the same configuration as a 115-volt outlet. After that, the percolator didn't perk much.

Figure 6-5 shows commonly used receptacle configurations. There are many special receptacles for radios, television antennas, public address systems, and many others. These configurations are not detailed in this book.

Lampholders

Lampholders (sockets), such as the kind used in ceiling and wall fixtures, and portable lamps such as table lamps can only be connected to a maximum 20-amp circuit. This does not concern someone with a single dwelling although it is an NEC requirement. A single receptacle installed on a branch circuit must have an ampere rating at least that of the overcurrent protection. Example: A 30A circuit must be supplied with a 30A receptacle. Continuous loads must not exceed 80 percent of the rating of the branch circuit overcurrent protection. A load defined as continuous is one that is expected to continue for three hours or more.

Feeders

Installation requirements of feeders are determined by calculations made of the demand load to be carried by them. Feeder conductors for a dwelling unit need not be any larger than the service-entrance conductors. Example: If the service-entrance conductors have an ampacity of 55 amps or less, the feeder conductors must have the same ampacity. In practice, feeders should not be less than No. 10 copper wire. A feeder consisting of two hot wires and one neutral wire, furnishing 240/120 volts, can supply two or three sets of three-wire feeders. When run in a metal raceway, be sure that all the wires of a circuit (one or two hot wires and neutral) are in one raceway. *Raceway* is defined as an enclosed channel designed expressly for holding wires, cables, or busbars. It may be of metal or insulating material. Conduit and Thinwall tubing are examples.

Branch Circuit Loads

General-use circuits of 15A capacity can have eight to ten outlets (receptacles or lighting fixtures) connected to them. Circuits of 20A capacity can have 13 outlets. Loads for additions

General Purpose Straight Blade Devices — NEMA Configurations

TYPE	VOLTAGE		15 AMP	20 AMP	30 AMP	50 AMP	60 AMP
2-POLE 2-WIRE	125V	1	1-15R †A2				
	250V	2		2-20R A2			
2-POLE 3-WIRE GROUNDING	125V	5	5-15R A5-11, F3, G3	5-20R A17-21, F5, G5	5-30R A28	5-50R A29	
	250V	6	6-15R A12-15, G4	6-20R A23-26, G7	6-30R A28	6-50R A30	
	277VAC	7	7-15R A16	7-20R A27	7-30R A29	7-50R A30	
3-POLE 3-WIRE	125/250V	10		10-20R A31	10-30R A32	10-50R A33	
	250V 3φ	11	11-15R	11-20R	11-30R	11-50R	
3-POLE 4-WIRE GROUNDING	125/250V	14	14-15R	14-20R A34	14-30R A34	14-50R A35	14-60R A36
	250V 3φ	15	15-15R	15-20R A34	15-30R A35	15-50R A36	15-60R A36
4-POLE 4-WIRE	120/208V 3φY	18	18-15R	18-20R A37	18-30R	18-50R	18-60R A37

NOTE:
For quick reference purposes, this chart shows only the female configurations applicable to receptacles, connectors and flanged outlets. Mating configurations for plugs and flanged inlets are shown on catalog pages indicated.

†Plugs that comply to NEMA 1-15P are listed on page A2.

HOW TO USE THIS CHART:

NEMA configuration ⟶ 2-20R

Catalog page where product is listed ⟶ A2

Fig. 6-5. General-purpose straight-blade devices (receptacles), NEMA configurations. (Courtesy Crouse-Hinds Co.).

to dwellings or additional wiring in dwellings must be based on 3 watts per square foot. Any one motor larger than ⅛ HP must be added to the load at 125 percent of the nameplate rating. In a dwelling, all general-use circuits should be divided so that if one circuit is out, the area fed by that circuit will not be in complete darkness. In any event, the load on any circuit may not exceed a total determined by the overcurrent device and by the ampacity of the wires protected by such overcurrent device.

There must be one special appliance circuit for each of the following: laundry, furnace, pump, garage door opener, and air conditioning. If electric space heating is planned, its load is calculated at 125 percent of the full load current (nameplate rating). This is to protect the insulation from overheating during long periods of continuous usage. See Table 220-11 of the NEC.

Fixed Appliance Load

Clothes dryers must be added to the total load as 5 kW each or the actual nameplate rating if higher. Where heating and air conditioning are both installed in a dwelling, only the air-conditioning load needs to be considered. Both loads will never be connected at the same time. Be sure to add in the load of the furnace fan to the air-conditioning load as the fan also runs for the air conditioning.

On a 240V, three-wire system, the neutral in the feeder and service-entrance cable carries only the *unbalance* current between the two hot legs on the 240V supply. This is illustrated back in Fig. 2-2. For electric ranges and other permanent electric cooking units, the maximum unbalance load is considered as 70 percent of the load on the ungrounded (hot) conductors. Therefore, a range needing No. 8 conductors can use a No. 10 neutral. When you use service-entrance cable, this arrangement of wire sizes is used in its construction.

Balancing the load is accomplished by arranging the circuits so that both hot legs of the 240V service carry nearly equal loads. Example: If the 120V circuits are three at 15 amps and five at 20 amps, connect two 15A circuits and two 20A circuits to the left leg of the distribution panel and one 15A and two 20A circuits to the right leg of the panel. The neutral will carry an unbalance amperage of 15 amp. This is only an example. The named circuits will almost never be fully loaded. This balance is close enough because the neutral has more than enough ampacity; later you might want to add more circuits. When connecting the circuits, maintain the balance as closely as possible considering the load of the original circuits.

Chapter 7

Concealed and Exposed Wiring

Electrical wiring takes many forms, but because much of it is concealed, we don't notice it. Wiring is concealed mainly for aesthetic reasons. Other reasons for concealing wiring are protection of the wiring itself and protection of persons who might come in contact with the wiring. Vandalism is especially a concern in public buildings. Exposed wiring is so called because the wiring is visible and it is mounted on walls rather than inside walls. The wire itself is not exposed. It is encased, first with insulation, then with either a sheath of tough material such as Romex has or a metal conduit protecting the wires. The old knob-and-tube wiring actually was exposed. It is now used in specialized areas and in industrial buildings.

CONCEALED WIRING

In a one-family dwelling, most wiring is inside walls supplying ceiling fixtures and wall boxes used for switches and receptacles. See Figs. 7-1, 7-2, and 7-3. Exposed wiring can be seen in basements and attics. The exposed wiring in basements, unfinished garages, and outbuildings must be protected from damage. This means that the cable must be run through holes bored in the center of studs or joists, especially where these are exposed and fastened every 4½ feet and anchored within 12 inches of a junction box. For plastic wall boxes without a metal clamp inside, the cable must be anchored within 8 inches of the box. See Figs. 7-4 and 7-5. Less protection is needed in attics except around access openings to the attic where this area is considered exposed, and, the cable must be anchored and taken through bored holes where flooring might be laid.

Where buildings have hollow walls, wiring is easily concealed inside these hollow spaces. Concealed wiring is most easily installed during construction of the building and before the drywall or other material is installed. First the

Fig. 7-1. Mock-up showing types of boxes and wiring methods. Top: 2-gang plastic switch box with Romex and BX run to it; ½-inch deep by 4-inch round ceiling box fed by Romex. Bottom: 4-inch octagon box on a bar hanger, fed by BX. Notice the bare wires are grounded to the two metal boxes.

wall and ceiling boxes are mounted in place, then the cable is run after the holes are bored in the framing members. At this time, no devices are installed. When the cable is installed, leave 6 to 8 inches extending from the wall or ceiling box. Then coil this end up and push it back into the box so that it will not interfere with the application of the wall finish.

This "rough work" must now be inspected by the governing authority. If the work is approved or after the work is corrected to the inspector's satisfaction, the wiring can be "closed in" or covered with the wall finish. No fixtures or other devices are installed until the wall surface is finished. At the time the rough work is being done, the service-entrance equipment and the service cable or conduit and meter base are installed. This type of construction is known as

Fig. 7-2. Close-up of the octagon box and the grounding screw. Box can be locked at any point on the bar and the bar is adjustable for length. Bars are available in two lengths. Notice special design BX clamps that hold and show the red insulating bushing required by the NEC.

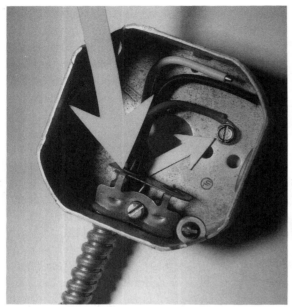

Fig. 7-3. A 4-inch octagon junction box for BX installation. Notice the special BX clamp that allows the red insulating bushing to be seen by the inspector.

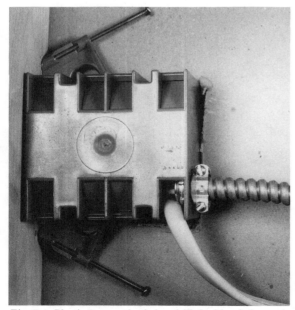

Fig. 7-4. Plastic 2-gang plastic box drilled with a hole saw to take a BX connector. Nails will be driven in all the way.

Fig. 7-5. Plastic wall box mounted to a stud by driving two screws through the box side and into stud. This is old work.

"new work" because the building is new or has not been finished on the inside.

If the building has the inside walls finished and wiring needs to be installed (such as an old farmhouse that never had electricity), it is considered "old work." See Fig. 7-6. The adding or extending of circuits in a new finished building is also old work. The words "old" and "new" are not a definition of age but a definition of the type of wiring method used. Wiring or rewiring of an older building is very hard, dirty work, but it is rewarding. Old work, being harder to do, takes longer. More cable and sometimes different types of wall boxes need

framing members. Protect Romex from nail penetration. Cover it with ⅟₁₆-inch metal plates where it is exposed to possible nail penetration.

Now is the time to plan for lighting fixtures and receptacles in the basement. Shallow boxes can be used if 1 × 2 furring strips are used on the basement. It might be necessary to chip out behind the wall box slightly. Boxes are available that are about 1⅛-inches deep. NOTE: Only one Romex cable is allowed in these shallow boxes. Drywall, ½ inch thick, combined with furring ¾ inch thick will accommodate this box for a switch or receptacle. The box must be flush with the finished wall surface or slightly

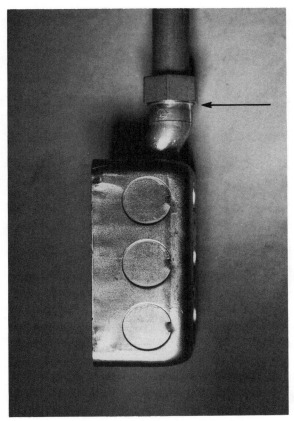

Fig. 7-6. An offset Thinwall connector designed to eliminate the bending of an offset in Thinwall. This saves much time.

to be used. Runs cannot be as direct as new work. Old work entails lots of fishing and hole drilling. Try to save wood trim if you have to remove it.

Basement Areas

Open areas in basements such as ceilings require neat work. Support the cable every 4½ feet. Secure it so as to prevent damage to it. Many basement ceilings are of the suspended type so cable does not have to be run through bored holes in joists. The basement walls need to have the cable run through bored holes in

Fig. 7-7. Metal wall box with ears, mounted on wood lathe, as found in older homes. This is a mock-up only, to show how the lath is cut and the box is mounted.

recessed. See Fig. 7-7. Any gaps must be patched with spackling compound so there is no opening between the box and the drywall.

Attic Areas

Wiring is not subject to great damage in attic areas, therefore, protection is only required within 6 feet of the attic access opening. If the attic is floored and used for storage, a light must be provided with a switch at the bottom of the stairs. The wiring must be protected just as in a basement area. See Fig. 7-8. Under the attic flooring, the cable must be run through holes in the joists. If a ladder must be used to gain access, the 6-foot rule applies. A light is not required, but it would be a good investment if the area is used for storage. I have one in my ladder-access attic. It is operated by a pull-chain on the lamp holder, but it is also wired, as the garage ceiling light is, by means of the same toggle switch. This way the light can only be turned on when the garage light is on, and hence a pilot light is not needed.

A garage is not required to have a disconnect; a toggle switch is sufficient. A toggle switch is approved for other residential outbuildings. Three-way switches are also approved. In one home I owned, I ran No. 10 wire to my garage and put a fuse panel in the garage. If a fuse blew, I would not have to go to the basement to replace it. My power tools were used in the garage.

Overhead Installations

Overhead wiring is still used, but great care must be taken to have the overhead wires high enough to clear any vehicle driven underneath. They must be securely anchored with approved insulators. Taps must be taken off and led down through a service head into a conduit and then

Fig. 7-8. Romex in notch of framing member and covered by a special ¹⁄₁₆-inch metal plate. This is to prevent nails or screws from being driven into the cable. This meets the requirements of Article 300-4(a) of the 1987 NEC.

into the building. See Fig. 14-36. Consideration must be given to:

▶ the span.
▶ the strength of the wires.
▶ the "sag" in hot weather.

No. 10 wire is the smallest size permitted by the NEC for overhead spans. The longer the span, the larger the wire size. If the outdoor temperature is in the range of 90 to 100 degrees Fahrenheit when the wires are strung, be sure to leave slack in each span. Allow at least 1 inch for short spans and 2 to 3 inches for long spans.

Knob-and-Tube Wiring

Knob-and-tube wiring is generally not used in dwellings. It is prohibited by the NEC except

is "knob-and-tube wiring" and should only be displayed in a museum.

EXISTING SERVICE-ENTRANCE EQUIPMENT

Service-entrance equipment that is already in place with intention to be rewired is usually not suitable for reuse. The equipment might have too small a capacity and perhaps be outdated. The original equipment probably had only fuses and the contacts in the fuse sockets might be burned or otherwise damaged. Replacement parts might be available, but it is better and probably less expensive to buy all new equipment. The old equipment will have served its purpose and has done the job for which it was installed. Remove it to make room for new, larger equipment. Quality equipment with extra capacity will last at least 20 years if it is properly installed and not overloaded.

If only two to four new circuits are needed, an additional fuse or breaker panel and cabinet can be purchased and connected to the main panel cabinet by means of an offset *nipple* or a *feeder* cable if the panel must be some distance from the main panel. If adding these additional circuits—with their added loads—will overload the service-entrance conductors, then there is a question of having adequate capacity of the complete service-entrance equipment.

In such a case, figure the total connected load and buy new service-entrance equipment and cable. New equipment will simplify the installation and make a neater job. Add-ons always *look* like add-ons. If necessary—because of basement remodeling work—the distribution panel can be installed remote from the service-entrance panel with its main breaker.

The NEC requires the main breaker to be as close as possible to the entry point of the service-entrance cable into the building. The ca-

Fig. 7-9. Method of mounting a ceiling box where a joist interferes with centering the box in the room. Notice that the joist has been notched. The box is ½ inch deep and is made for these situations. Notice the Romex clamp and the hex head grounding screw securing the ground wire.

in industrial and agricultural installations. Knob-and-tube wiring concealed within the hollow spaces of building walls is severely restricted by the NEC. I do not recommend adding on to or repairing old knob-and-tube wiring. It is better and safer to remove it and install new cable wiring in its place. When abandoning such wiring, make sure there is no way that it can be reconnected to any new wiring. These old systems are a fire hazard. I have touched such old wiring and the insulation has fallen off in my hands. Any wiring consisting of single wires supported on porcelain knobs and going through porcelain tubes inserted through studs and joists

ble on the street side of the main breaker has no overcurrent protection and should be on the outside of the building, except that portion necessary to enter and connect to the panel of the main breaker.

Equipment over five years old and newer equipment showing signs of arcing or burning should not be reused. Panels using fuses sometimes show signs of deterioration of the fiber washer underneath the center contact of the fuseholder. If the panel is in a very damp location, these fiber washers become water soaked and will pass electricity (actually bypassing the fuse entirely). Other, old installations will have fuses in both wires (the live wire and the grounded wire). This is prohibited by the NEC. These prohibited conditions should be first priority in making repairs and changes, as it is imperative that they be corrected.

SERVICE-ENTRANCE CABLE (SE)

Service-entrance cable is similar to Romex except that the wire is type XHHW (moisture- and heat-resistant insulation). There are two insulated wires (red and black). The neutral wire consists of many fine strands of tinned copper wire wrapped spirally around both insulated wires. This is then covered with a tough outer covering that is very damage resistant. There is no grounding wire as such. The neutral is also the ground wire. This cable brings power through the meter and into the building to the service-entrance equipment. The fine tinned strands of the neutral are bunched and twisted together and connected to the neutral bar and to the grounding wire in the panel and in the meter base.

Modern homes usually need large service-entrance cable wire size such as No. 4, 3, 2, or 1 for 100A, 115A, or 130A capacity respectively. Service-entrance cable can also be used to wire ranges and clothes dryers using Nos. 6 and 8 respectively. Watertight cable connectors must be used in wet locations such as the top entrance to the meter base outdoors to prevent moisture entry. A special service head is used at the point where the service drop connects to the cable. You can purchase service-entrance cable by the foot at home improvement centers. Measure carefully to be sure that you have enough length. This cable is very expensive. Cable with aluminum wires is also satisfactory for ranges, dryers, and electric furnace uses.

CONDUITS

Conduits take different forms and they are made of different materials. The following sections describe them in the order that they were developed.

Rigid Metal Conduit. Rigid metal conduit has been in use almost as long as wiring has needed protection from damage. This conduit is similar to water pipe in appearance, but it is specially treated to make the insides smooth and to bend easily. It is threaded similar to water pipe, but it should be cut with a hacksaw rather than a pipe cutter. A cutter will raise a large burr on the inside, and the burr must be completely removed with a reamer. Because this conduit requires a bender, vise, and pipe dies, the homeowner might not want to use it. It is readily available at electrical wholesalers. Fittings that eliminate threading are also available. Certain sizes now come made of aluminum.

Electrical Metallic Tubing (EMT). This tubing, commonly known as Thinwall, is easy to bend and it can be cut with a hacksaw. It is

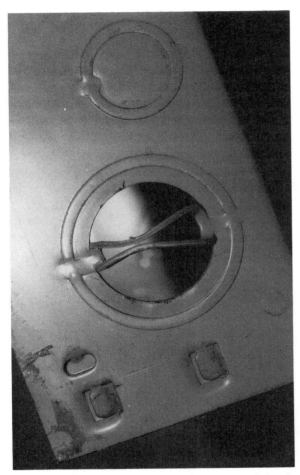

▶ Crimp-on that needs a large crimp-on tool.

▶ Compression type using a nut and split ring to tighten onto the Thinwall.

▶ Setscrew type. The setscrew types are either steel or die cast. These have one or more setscrews depending on the size of the fitting.

▶ For use with internally threaded hubs on cast boxes and meter hubs, a Thinwall adapter slips over the Thinwall end, and when tightened into the hub grips the Thinwall tightly. See Figs. 3-6 and 3-7.

Rigid Nonmetallic Tubing. This type of conduit is similar to plastic water pipe except that it is made especially for wiring use and can be used underground. Common sizes are of polyvinyl chloride. While metal Thinwall needs supports only every 10 feet, the nonmetallic type needs support every 3 feet in the ½-inch to 1-inch sizes because it is more flexible. Because this conduit needs specialized equipment to bend it, the homeowner might find that it is

Fig. 7-10. Method of removing sections of a concentric knockout. Center round knockout is first removed, then sides of next ring are pried up together and bent back and forth until they break away. Remaining pieces must be hacksawed away at times. Do not loosen remaining sections as they must remain solid to give good grounding to fittings used.

bent with a Thinwall bender that can be rented. The bender sells for under $20 and it will pay for itself if you plan to do much work with Thinwall. This type of conduit is never threaded. It is connected to boxes with Thinwall connectors, and couplings join lengths together. See Fig. 7-11. These fittings are made in various types:

Fig. 7-11. Method of tightening a Thinwall connector to a wall box. Inside locknut is snugged up and the outer part of the connector is tightened while holding the inside locknut.

77

not practical for home use, however formed 90° curves are available.

Flexible Metal Conduit (greenfield). This conduit appears to be the same as BX. The steel armor is the same, but the inside is empty. Greenfield is usually larger than BX. The ½-inch size is larger than ½-inch Thinwall in outside diameter. For some areas, the ⅜-inch size is approved for lengths under 6 feet for connecting to light fixtures. BX connectors are used for ⅜-inch greenfield conduit. Larger sizes use greenfield connectors.

Liquid-tight Flexible Conduit. This type of flexible conduit is used commercially for connections to motors, etc., where dampness is present. Special fittings are used to maintain the dampness resistance. The name Sealtite identifies a metallic type. Plastic types are also manufactured.

Chapter 8

Grounding: Theory and Importance

Grounding plays a very important role in the safety of any electrical system—whether it is in a garage or a skyscraper. Read through the following list of terms, as I refer to them throughout this chapter.

continuous ground—All circuits must have a grounding wire connecting all metal (noncurrent-carrying) parts, cabinets, wall and ceiling boxes, fixture canopies, and mounting brackets holding toggle switches and receptacles in their respective wall boxes. This system, if carefully installed, is a direct connection to ground for each device bracket, fixture, box, and cabinet.

green wire—Extension cords, portable saw and drill cords, and other cords having three wires; having a three-blade attachment plug designates them grounding cords. The cord is made up of a rubber (or plastic) covering, jute fillers to make a round cord, and three wires—one black wire (hot), one white (neutral) and one green. This green wire is the *safety* wire. Without this wire, any tool or appliance would be a potential hazard to the user. *Never* cut off the third grounding prong from a three-wire attachment plug.

ground—(NEC designation: *grounding electrode*). The pipe of an underground metal water system. A *made electrode* is driven pipe or ground rod used where there is no underground piping system.

ground, to—Connecting a wire (grounding wire) from the metal parts (that are noncurrent carrying) such as cabinets or neutral wire to the ground.

grounded neutral wire—This is the white or natural gray color wire coming in to the service-entrance panel. If service-entrance *cable* is used, the neutral wire will be bare strands twisted to make a wire shape.

grounding wire—This wire does not normally

carry current. It can be bare or have green-colored insulation. It is connected to the metal parts of the system such as cabinets and frames of home appliances, fuse, and breaker enclosures. The only time current flows in this wire is when a fault occurs in the current-carrying parts of the equipment. This grounding wire must be installed and maintained in tiptop condition with low resistance so that the flow of fault current will be great enough to blow the fuse or trip the breaker promptly.

ground wire—(NEC designation: *grounding electrode conductor*). A wire connected to the neutral terminal bar in the service-entrance panel and also to the metal cabinet itself, plus the conduit or BX or the ground wire in Romex cable.

insulation—The neutral wire must be insulated (except in service-entrance cable) and treated the same as a hot wire. This is a Code requirement.

neutral integrity—The neutral wire must never be disconnected by a fuse, breaker, switch, or other device. If such a device can break the hot wire *and* the neutral simultaneously, it can be used. This is extremely important. The neutral wire must run without a break from the service-entrance panel to the place where the current is consumed. For 120V circuits, there are no exceptions.

nonuse of neutral—Some appliances and motors rated at 240 volts do not need a neutral wire. A separate grounding wire is used for equipment grounding unless conduit runs to the equipment. Conduit acts as the grounding wire.

white neutral wire—Large-size insulated wires do not come in white. In this case, both ends that are visible are painted white or taped with white tape so that the wire is iden-

tified as the neutral wire. This then is the neutral wire.

ADVANTAGES OF GROUNDING

While grounding is required by the NEC, it is very much to your and others' advantage to maintain a properly installed ground at all times. If a high voltage line of, say, 4800 volts should fall across your 120/240V service in a storm, a serious condition would exist. This and lightning strikes are very greatly minimized by a good solid ground wire connected to a proper ground. See Fig. 8-1.

Assume a motor has an internal fault (a bare hot wire inside the motor housing touches the housing). If the motor housing is grounded, the fuse will blow or the breaker will trip. If the housing is not grounded, you will get a shock if you touch the housing. This is especially crucial when using power tools outdoors. Be sure they are grounded properly.

If grounding is not done properly, such as having the fuse in the neutral instead of the hot wire or fuses in both hot and neutral wires, severe shock or even death could occur. In some installations, the disconnect switch opens the neutral, leaving the hot wire connected. In this case, the equipment is still hot even though it is not operating. As with all electrical work, attention to detail and exactness in following the NEC and checking the completed work for defects and mistakes is the ultimate goal of the good electrician.

OUTBUILDINGS

Outbuildings supplied from a main building must have their own service-entrance equipment provided with a separate grounding electrode with a grounding conductor. See Fig. 8-2. There are exceptions to this rule, but for farm

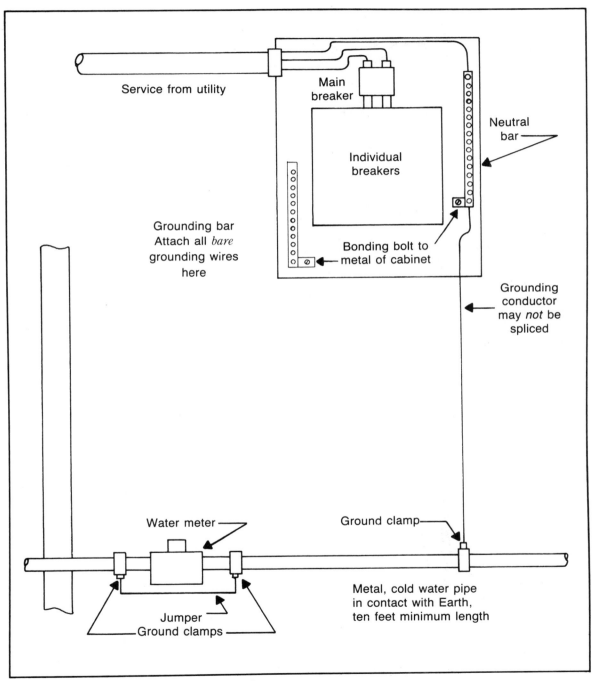

Service from utility

Main breaker

Individual breakers

Neutral bar

Grounding bar Attach all *bare* grounding wires here

Bonding bolt to metal of cabinet

Grounding conductor may *not* be spliced

Water meter

Ground clamp

Jumper
Ground clamps

Metal, cold water pipe in contact with Earth, ten feet minimum length

Fig. 8-1. Layout for the grounding system for a dwelling.

81

Fig. 8-2. 60A service-entrance panel. The bare neutral is connected to the neutral bar at the top of the panel. Range cable leaves box at top right. Its neutral also connects to the neutral bar as does the ground wire to ground the system. The ground wire must be grounded to a metal cold water pipe of which ten feet must be in contact with the earth. The neutral bar extends down behind the panel itself and has terminals (bottom center) for the white wires of the individual circuits. Notice the method of grounding to the cold water pipe at the right.

buildings I advise that each outbuilding entrance panel be grounded. The cost is small when done as the building is being wired. The grounding *electrode* must be as near as possible and in the same area as the grounding *conductor* from the system. Grounding electrodes should be:

▶ The nearest available effectively grounded steel structural member of the building.
▶ Or the nearest available effectively grounded metal cold water pipe (the water pipe must be buried in the earth for at least 10 feet).

▶ Or other electrodes specified by NEC Sections 250-81 and 250-83 where the first two electrodes are not available. The *made electrode* described in code Section 250-83 is used on farms and cottages where other electrodes are not available.

Rod or pipe electrodes are commonly used. They are to be driven vertically 8 feet into the ground. If rock is encountered, they may be driven at an angle. Eight feet is the minimum; sometimes the driven ground is required to be driven more than the 8 feet so that the upper end is recessed below the normal ground level. The grounding conductor is then clamped to the top of the electrode with the grounding clamp. After inspection, the electrode is then covered with earth for protection.

Plate electrodes must be at least two square feet in area. Ferrous plates must be ¼-inch thick; nonferrous plates must be a minimum of 0.06 inches thick. I recommend a greater thickness of the nonferrous plates because parts of such a thin plate could break off and reduce the effective surface area, impairing the grounding system.

NEC Table 250-94 lists grounding electrode conductors sized for the largest wire in the entrance cable. For dwellings, the minimum is No. 8. The grounding conductor must be one continuous length. Splices are never permitted. The conductor can be of copper, aluminum, or copper-clad aluminum. This can be exposed if it is not liable to be damaged. Staple it to wood or use clips with anchors on concrete. It is not recommended you run it inside conduit or greenfield, because special conditions cause stray eddy currents to be present from having the wire inside metal conduit. Nonmetallic conduit could be used for protection in exposed places. There are special ground clamps available to accommodate conduit and greenfield, plus the grounding electrode conductor. For the home and farm, you very seldom have need for mechanical protection. If you do need protection, consult an electrical inspector before using metal conduit.

Equipment-grounding conductors can be of copper, aluminum or other corrosion-resistant metal that is stranded, solid, insulated, or bare. Conduit can be used as the grounding conductor. The bare ground wire in Romex, BX, and the green conductor are to be used as grounding conductors. The correct installation of the equipment grounding conductor is very important.

Every connection of the equipment grounding system must be mechanically and electrically tight (this means that an electrical fault must be able to travel all the way back to the service-entrance panel and then to the ground electrode with very low resistance so that the fuse or breaker opens). High resistance might not allow the overcurrent protection to disconnect the equipment (or person) from the hot wire. This means that the metal housing of the electric drill motor is directly connected to the water pipe or other ground. This protects you from shock, providing the cord and plug are in perfect condition and the receptacle is properly grounded. This is why you are warned never to cut off the grounding blade on the attachment plug.

Electric clothes dryers and electric ranges are permitted to have the noncurrent-carrying parts (stove-top, oven door and sides) grounded by the system neutral. Service-entrance cable can be used and the neutral can be bare *only* if the cable originates at the service-entrance equipment panel. The NEC Section 250-60 in its entirety and Section 250-61, Exception No. 1 detail this requirement.

Duplex receptacles of the grounding type must have solid metal-to-metal contact with the metal box in which they are mounted. Surface-mounted boxes such as ''handy'' boxes give tight contact. Wall boxes sometimes are recessed below the finished wall surface and do not have tight contact. The plaster ears holding the box flush with the wall prevent this. Code Section 250-74, Exception 2, allows a special type of yoke to be used (''Underwriters listed''). A special spring holds the screw tightly. All devices do not have this; if not, use a jumper wire from the device green ground screw to the metal box and the grounding conductor in the cable or BX. Don't forget to make this connection.

Where an ungrounded receptacle is to be replaced by a grounded receptacle, the green grounding screw must be connected to the metal wall box with a green hex head grounding screw. If this metal box is not grounded (check with your voltage tester), the green screw must be connected to a cold-water pipe. If this is impossible, a new ungrounded receptacle must be used (one that will accept only two-prong plugs). A grounded-type receptacle that is not grounded will give a false sense of safety and lead to serious consequences! *Note:* A ground fault circuit interrupter (GFCI) can be used as a replacement receptacle (NEC Article 210-7(d)), provided the grounding wire of the GFCI is not connected to any wire but has the bare end wrapped with tape and is not used. See Fig. 8-3.

Short sections of raceway (conduit, etc.) must be grounded. This might be a place where Romex is brought down a concrete basement wall for an outlet or switch. As the Romex will have the grounding conductor at the point of attachment to the device, merely attach the grounding conductor to the green box-grounding screw. Screw the device mounting bracket to the handy box ears and the device is effectively grounded.

THREE-TO-TWO PLUG-IN ADAPTERS

Plug-in adapters are sold everywhere, and they are used to supposedly convert the two-slot receptacle for use of the three-blade attachment plugs found on grounded power tools and appliances. They either have a spade terminal

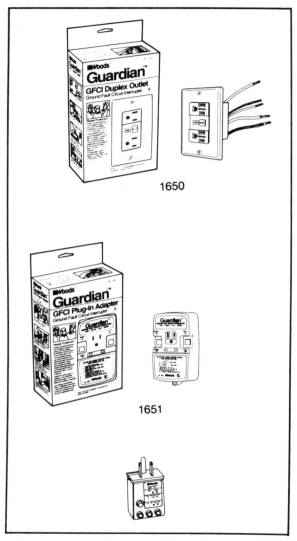

1650

1651

Fig. 8-3. Two styles of ground fault circuit interrupters (GFCI's), permanent, wired-in, and a "plug-in" style. The plug-in is inserted in a receptacle and secured with the screw holding the plate. Also shown is a GFCI Tester/Circuit Analyzer which also tests and standard grounded or non-grounded outlet to detect faulty wiring. (Courtesy Woods Wire Products, Inc.)

on a short pigtail or an eyelet. Either connection is fastened under the center plate screw on a duplex receptacle cover. Unless it is determined that this center screw is in fact providing a good positive ground all the way back to the system ground, there is no assurance of safety. Everything from the service-entrance panel and throughout the dwelling must be re-wired using Romex with ground to provide properly grounded receptacles. See Fig. 8-4.

BONDING TO OTHER SYSTEMS

There are many reasons for bonding other grounded systems to the dwelling electrical ground. Interconnection is required for lightning rod systems, communications systems, and cable TV systems. Lack of interconnection can

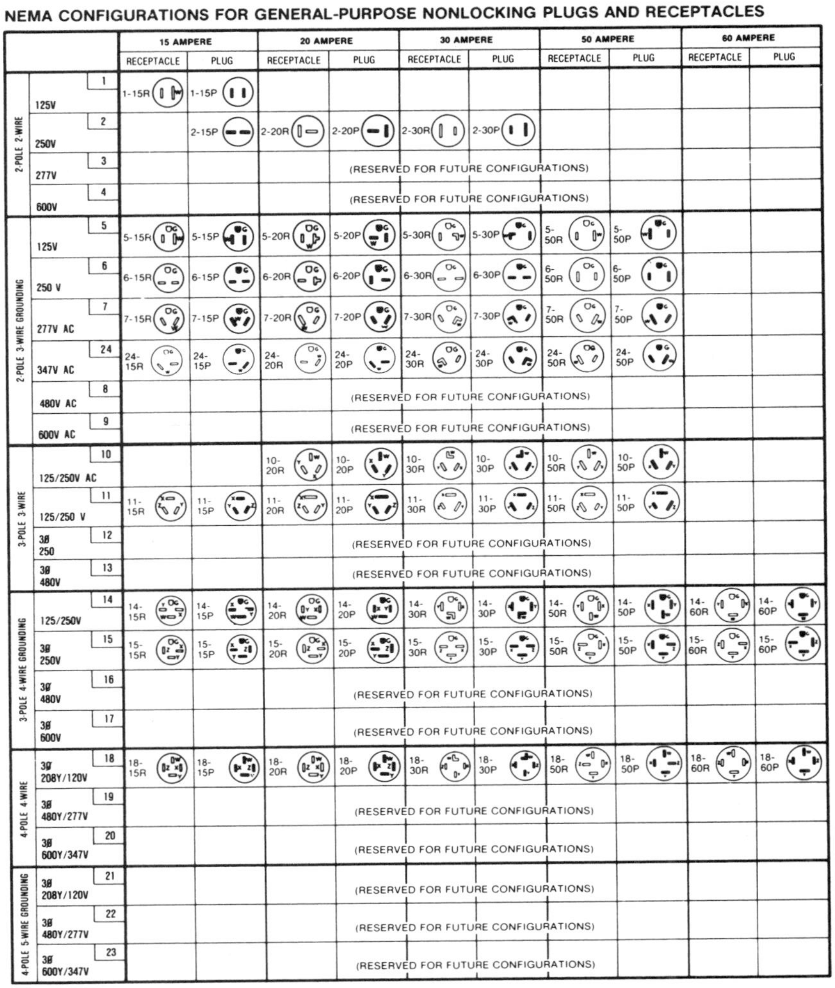

Fig. 8-4. Commonly used configurations for receptacles. These match plugs except in reverse. Note: Two-pole, 2-wire receptacles shown on the first two lines are obsolete and may be used only for replacement of other 2-wire receptacles. The NEC now requires that all current carrying devices be grounded (Courtesy Leviton Mfg. Co.).

cause severe shock and fire hazard. Cable TV cables have a metal casing, and lightning striking the cable can be led directly into the dwelling through the cable connection at the TV set. Many times there is so much plastic plumbing in a dwelling that it is hard to locate a true system ground. New installations are now providing externally mounted pressure-type connectors on the service-entrance equipment cabinet just for this interconnection.

METHODS OF BONDING SERVICE EQUIPMENT

Pressure-type connectors have a hole to bolt to the metal cabinet. At the other end is an opening for insertion of the solid or stranded grounding conductor. A setscrew tightens down on this wire and makes an excellent bond.

All of these methods are used to effectively ground the service cabinets and the service neutral of the incoming service. One type is a grounding bushing to bond the conduit. There is a pressure connector to accept the wire, and in addition there are two setscrews that go through the bushing and bite into the metal of the cabinet wall. Grounding wedges provide a clamping action and also have setscrews to bite into the bushing inside the cabinet, at the same time pulling up against the conduit threads. Rigid conduit screwed into the threaded boss of, say, a meterbase must be "wrench tight" so as to make a positive ground through the threads.

GROUNDING ELECTRODE SYSTEM

To form a grounding electrode system, the following items must be bonded together (if they are present):

▶ Metal underground water pipe.

▶ Metal frame of building if it is effectively grounded.

These must be bonded together with No. 6 copper wire the same as the grounding conductor. Solderless pressure connectors must be used to connect the grounding electrodes. Soldering is not allowed. The reason for this elaborate grounding electrode system is that metal water pipes are sometimes replaced with plastic pipes, thus losing the continuity of grounding. Other building alterations are made that can loosen or disconnect the grounding conductor. This is a dangerous condition. One used home I purchased had no grounding electrode conductor of any kind. I installed one very soon after buying the house.

When installing switches or receptacles in wall boxes—if there are two or more grounding wires entering the box—they must be twisted together, with a short pigtail, and then a wirenut is screwed on. The pigtail goes to the device mounted in the box. In this manner, if the device is ever removed for any reason, there will be a continuation of grounding. See Fig. 8-5.

A new type wirenut (green color) is available that has a hole through the closed end. Instead of as above, one grounding wire is extended through this hole and is connected to the device grounding screw.

Metal wall boxes also must be grounded by the ground wire. Fit a green grounding hex-head screw in the tapped hole in the box and clamp the ground wire underneath it. Nonmetallic boxes cannot be so grounded. Be sure to attach the cable ground wire to the green device screw. Approved ground clamps must be used to connect the grounding conductor to the grounding electrode (water or other pipe). Many different clamps are available for the

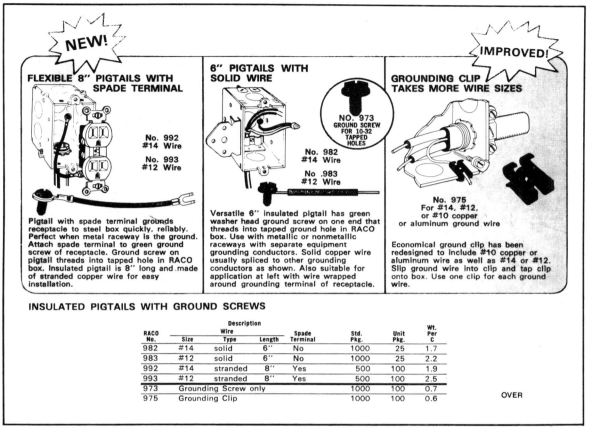

Fig. 8-5. Grounding devices to provide continuity of ground and to ensure safety. (Courtesy Raco, Inc.).

FLEXIBLE 8" PIGTAILS WITH SPADE TERMINAL

No. 992 #14 Wire

No. 993 #12 Wire

Pigtail with spade terminal grounds receptacle to steel box quickly, reliably. Perfect when metal raceway is the ground. Attach spade terminal to green ground screw of receptacle. Ground screw on pigtail threads into tapped hole in RACO box. Insulated pigtail is 8" long and made of stranded copper wire for easy installation.

6" PIGTAILS WITH SOLID WIRE

NO. 973 GROUND SCREW FOR 10-32 TAPPED HOLES

No. 982 #14 Wire

No .983 #12 Wire

Versatile 6" insulated pigtail has green washer head ground screw on one end that threads into tapped ground hole in RACO box. Use with metallic or nonmetallic raceways with separate equipment grounding conductors. Solid copper wire usually spliced to other grounding conductors as shown. Also suitable for application at left with wire wrapped around grounding terminal of receptacle.

GROUNDING CLIP TAKES MORE WIRE SIZES

No. 975 For #14, #12, or #10 copper or aluminum ground wire

Economical ground clip has been redesigned to include #10 copper or aluminum wire as well as #14 or #12. Slip ground wire into clip and tap clip onto box. Use one clip for each ground wire.

INSULATED PIGTAILS WITH GROUND SCREWS

| RACO No. | Description Wire | | | Spade Terminal | Std. Pkg. | Unit Pkg. | Wt. Per C |
	Size	Type	Length				
982	#14	solid	6"	No	1000	25	1.7
983	#12	solid	6"	No	1000	25	2.2
992	#14	stranded	8"	Yes	500	100	1.9
993	#12	stranded	8"	Yes	500	100	2.5
973	Grounding Screw only				1000	100	0.7
975	Grounding Clip				1000	100	0.6

OVER

various electrodes used. The surface of the electrode must be absolutely clean (use emery cloth to shine it). It must have no paint or corrosion that might prevent good contact. Watch for aluminum or silver paint or galvanized pipe at the point of clamping. Scrape this off as the pipe has to be absolutely bare.

Chapter 9

Wiring Methods and Techniques

Over the years, electricians have developed certain construction methods and techniques that enable them to do better work faster and more accurately. These methods are passed on to apprentices and helpers in training programs in the industry and by journeymen providing help and advice on the job.

EXTENDING WIRING IN BUILDINGS

When an additional receptacle is needed, many times power for it can be obtained from a nearby receptacle, provided that the circuit supplying that receptacle is not overloaded. If that circuit is carrying its full load, another circuit must be used or a new one can be provided by adding another breaker or fuse to the panel.

Assuming that you have made this determination and that the circuit can handle the extra load, this circuit can be extended. If you have located a junction box or light in the basement or have a nearby receptacle, proceed from this point. The basement light must not be switch-controlled. A pull-chain fixture is satisfactory. Connecting to a switch-controlled device would shut off your new receptacle.

When you have decided on the location for the new receptacle, remove the base shoe molding and drill a ¼-inch hole down at an angle through the corner between the floor and the baseboard. This will locate the exact point for drilling up from the basement—directly into the hollow wall space. Figure 9-1 illustrates this. Use a ⅝-inch or ¾-inch spade bit for this (using an electric drill motor). If you plan to do much of this type of work, it would be advisable to buy an electricians' 18-inch long, ¼-inch or ⅜-inch size wood bit for drilling pilot and test holes.

This preliminary work is done to determine if the wall space is hollow and will accept a wall box. Many electricians use a long spike to drive

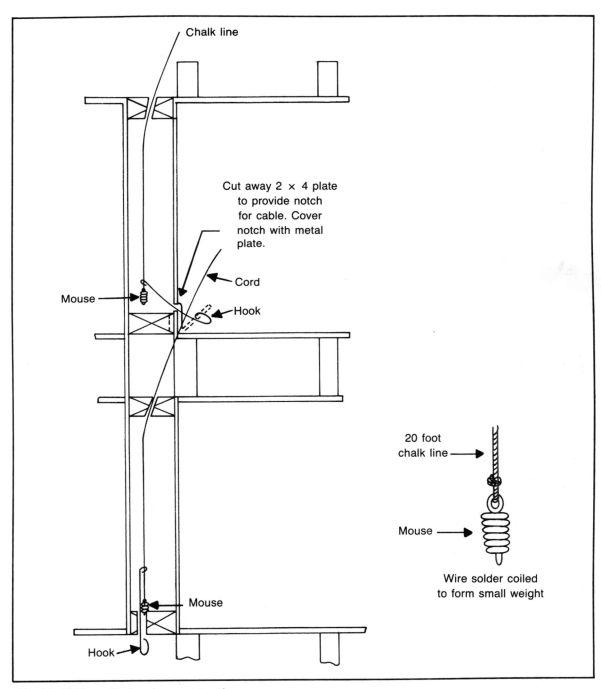

Fig. 9-1. Fishing cable from basement to attic.

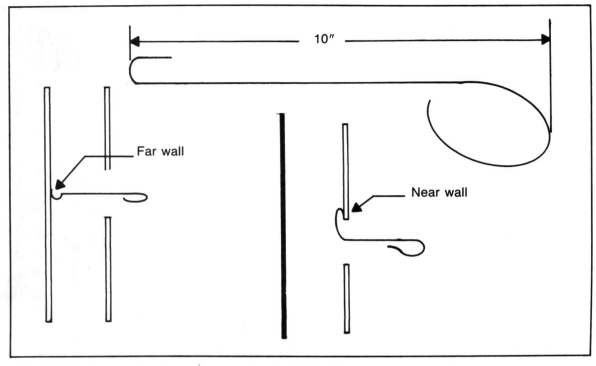

Fig. 9-2. Fishing tool used to estimate measurements and check for interference.

down at an angle without removing the base molding. If you first cut an opening in the drywall or plaster and then find that you cannot get cable to that location or that interference in the hollow wall prevents a box from being installed, you simply have to repair the wall.

Checking for interference or routing cable is known as "fishing." See Fig. 9-2.

To search for interference, straighten a wire coat hanger. Loop one end for a handle and probe up inside the hollow space beyond the height of the proposed box location. For switch box locations, two coat hangers must be "spliced" together or a longer stiff wire can be used. Switch box heights are standard at 48 inches. If you find that the wall will accept the box and cable, hold the box face up to the wall at the proper place and mark around all four sides. Trace around any edge projections on the box. Using a small screwdriver, punch a small hole in the center of the outline. Then wiggle the screwdriver sideways to see if there are any studs in the way. You can also use an L-shaped stiff wire to probe. Insert it past the bend of the wire and rotate the end to sweep the inside of the drywall or plaster to detect studs.

To cut out the opening, use a keyhole saw. You can also use a hacksaw blade with one end wrapped with tape to protect your hands. Coarse tooth blades are best. Gently start sawing at one corner while holding the blade almost parallel to the wall. This way you will be able to go through the drywall without first having to make a starting hole. Saw all four sides *almost*

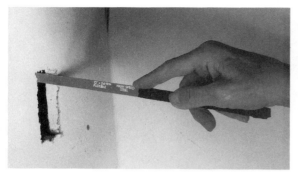

Fig. 9-3. Trimming an opening for a wall box to give clearance for insertion of the box.

all the way. Leave a little on each side. This prevents breaking the drywall in the wrong places. Then finish sawing each side. See Fig. 9-3.

When the opening is finished, try the box. Some trimming might be needed. Trim carefully around the box projections so as not to enlarge the hole too much. If the hole is too big,

Fig. 9-4. Wall box and two switches in one body. Voids around box are filled with spackling compound.

Fig. 9-5. Toggle bolts used to mount device on wall surface. These bolts can be used also to mount devices on hollow cement block walls.

you must plaster around it to seal the box. This is an NEC requirement. See Fig. 9-4.

Depending on the method of support for the box, you need to provide clearance for these supports. Some boxes have screws inside, which when tightened, spread arms out behind the drywall like toggle bolts to hold the box (see Fig. 9-5). Others have a U-shaped bracket that fits around the box and it is held by a screw through the box back (see Fig. 9-6). When the box and bracket are pushed into the wall opening (Fig. 9-7), the bracket spreads. When the

91

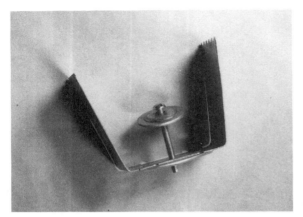

Fig. 9-6. Close-up of spring type box support. Bolt goes through the back of the box.

Fig. 9-8. Wall box in place. Spring-type box holder has been tightened and box is solidly fastened. This is a mock-up.

screw is tightened, it bites into the back side of the drywall, again like a toggle bolt (see Figs. 9-8 and 9-9).

The last type is sheet metal in the shape of the Greek letter Pi, called a Madison support having one long arm with two short legs at right angles near the center. See Fig. 9-10. The long arm is inserted into the wall opening vertically if the opening is vertical and the short arms are bent slightly to the side. One bracket is needed for each side of the box. The box is

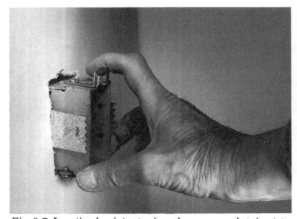

Fig. 9-7. Inserting box into opening, shows ears and spring-type box holder.

Fig. 9-9. Metal wall box with spring-type support mounted on a clear plastic mock-up wall instead of drywall. Notice the box brackets on the front side for a solid job.

Fig. 9-10. Set of Madison box supports.

ier to insert into the wall opening after the cable has been inserted into the box.

If the hole into the basement is large enough, the cable will slide back down when you insert the box into the drywall opening. Depending on the box anchoring device you are using, secure the box in the drywall. When you prepare the BX cable, remove 6 to 8 inches of armor from the end. (See Figs. 9-24 through

then inserted into the opening. After the box is aligned, the short arms are bent around and into each side of the interior and tightened against the box sides, out of the way of any device to be installed in the box. See Fig. 9-11.

When extending a circuit, it is best to install the new wiring backwards from the far end of the extension and make the final tie-in last. This method eliminates shutting off the power for a long time while you are doing all of the new wiring installation.

With the box opening cut in the wall and the hole bored up through the bottom or down from the top of the hollow wall, it is easy to shove the cable into the wall space and bring it out the opening. When buying a metal wall box, try to get one that has the two rear corners made at an angle. This will make the box eas-

Fig. 9-11. Metal box supported by a Madison box support. This is a mock-up using clear plastic as the drywall to show how the supports are used. The top arm has been bent in fully, the lower one only partially.

9-26.) Do the same if you are using Romex. Bend the wires so that they do not drop out of the box after the box is secured. If more than one receptacle is being installed, two knockouts must be removed. In addition, two holes are needed through into the basement unless the one is large enough for two cables. More instructions for working with BX and Romex are given in Chapter 14.

If two receptacles are to be installed back-to-back, do not place them exactly back to back because the wall thickness might not be enough to accommodate them in this position due to the cables. Try to leave 6 inches of space (horizontally) between the boxes where their backs are adjacent. When there is a stud, a hole can be bored through it and a cable can be pushed through. Another receptacle can be installed farther along the wall on the other side if necessary. Probe this next hollow wall space before cutting into the drywall. Figures 9-12 through 9-16 illustrate these techniques.

If you are wiring a switch or other wall-mounted device, you might need to shove a long piece of stiff wire up inside the wall space (or two straightened coat hangers) to reach up to the wall box opening. Working down from an unfinished attic is more difficult because you must be careful not to step on the plaster ceiling. Step only on the joists or boards laid across them.

Sometimes measurement is the only way

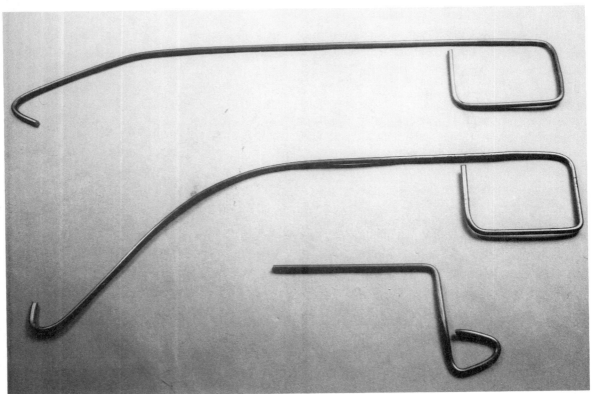

Fig. 9-12. Two shapes of "fishing hooks" (from wire coathangers) and one "stud finder" (to determine clearance inside wall).

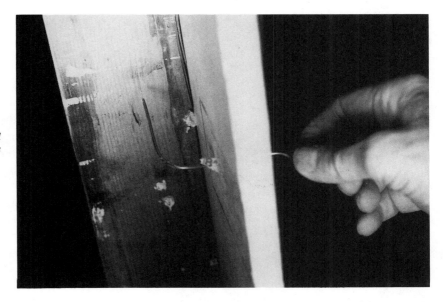

Fig. 9-13. Stud finder in use. Shows if wall box will clear the stud before cutting a large hole in the drywall.

to find access to the desired hollow wall space. Measure very carefully. I have made errors in measurement and drilled through ceilings instead of into the top of the hollow wall space.

Find a reference point that shows on the room ceiling and in the attic. A light fixture will show in both places. When you measure, take the center of both the fixture and the box in the at-

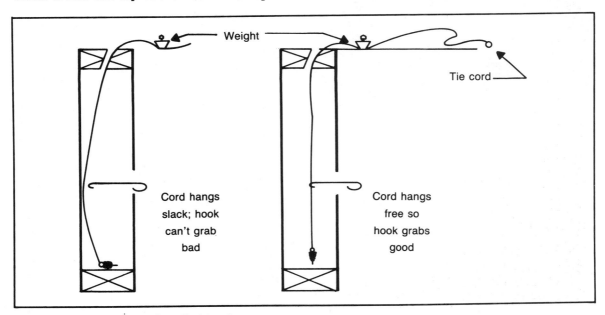

Fig. 9-14. Method of fishing in walls from above.

Fig. 9-15. Chalkline has slack here and will be difficult to hook.

devices (of various manufacturers) can be installed so that up to three devices—such as switches, pilot lights and outlets—will fit in a one-gang wall box. I would rather not use these devices because of the crowding of wires in a one- or two-gang box and the closeness of the terminal screws in the assemblies. They are approved, but use them only when necessary.

Second-floor finished rooms become a greater problem when adding wiring circuits. It might be possible to go toward the eaves through storage-space entrances. A flashlight aimed between joist spaces should provide clues to ways of running wiring underneath finished second-story flooring. See Figs. 9-18 and 9-19.

Where it is absolutely necessary to go all the way to the eaves and down an outside wall with cable, it might be simpler to notch the top

Fig. 9-16. Fishing tool can hook line because it hangs free (does not lie against the drywall).

tic, because an error of even 2 inches will bring the drilling into the room ceiling rather than in the hollow wall space. When going into the attic, keep your sense of direction so that you don't measure wrong.

Many metal wall boxes are made with removable sides so that two or more can be *ganged* together to accommodate more devices. Not all boxes are made this way. Plastic boxes are now made in two- and three-gang style (see Fig. 9-17). The interchangeable line of wiring

wall plate after cutting through the plaster corner at the junction of the wall with the ceiling. When flooring has to be removed on the second floor, it is a painstaking process.

If the boards are tongue-and-groove, one board must have the tongue split off so that board can be pried up and removed. Use a sharp stiff putty knife as a chisel. Drill a series of ⅛-inch holes to allow the saw blade to be started. Saw the board at each end, using a keyhole saw. The board should span three joists instead of two for better strength after replacement. When you are using the putty knife to split off the tongue, you can then locate the joists. If you use a power circular saw, set the blade to cut through just the floorboard thickness and center the cut on the center of the joist. This eliminates having to nail support cleats on the sides of the joists to support the ends of the removed floorboards.

Save the sawdust to make a matching filler for the saw cuts after replacing the flooring. Mix with glue to make filler putty. If hardwood flooring has to be removed to run the cable, try to remove flooring in an inconspicuous area such as a closet rather than in an open hallway or bedroom.

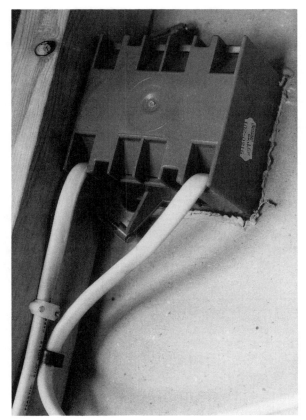

Fig. 9-17. Two-gang plastic wall box nailed to stud. Romex cables have been stapled within 8 inches of box. This is a mock-up showing "new work" installation.

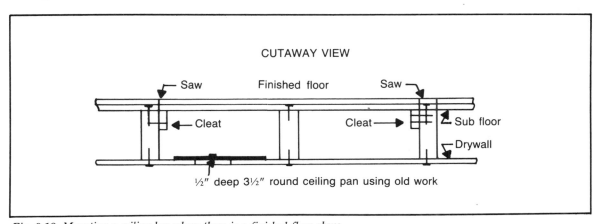

Fig. 9-18. Mounting a ceiling box where there is a finished floor above.

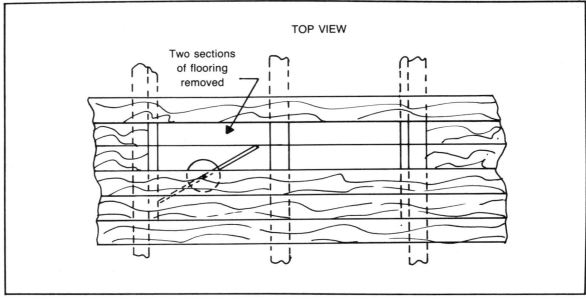

Fig. 9-19. How to remove and replace second-story floor for wiring a first-floor ceiling outlet.

Figure 9-1 shows how a baseboard can be removed and cable run behind it where there is no other route. This method runs the cable in notches cut in the studs. When using non-metallic cable (Romex), protection must be provided to guard against driving nails through the cable when nailing back the baseboard. Two methods are approved:

▶ Nonmetallic cable protectors: 16-gauge metal plates with prongs for pounding over the framing notches after the cable is run.
▶ Thinwall tubing (EMT) is set in notches cut in the studs; run the cable through for protection. The ½-inch size is usually large enough for 14-2 or 12-2 cable.

Many older houses have lath and plaster rather than the typical drywall. When cutting out for wall boxes, locate the box so that it is centered on one lath. This leaves a part of the top and bottom lath to screw the box ears in place. See Fig. 9-20.

Mark the box outline on the plaster. Remove all plaster from the laths inside the box outline. Be careful not to loosen the lath from the back side of the plaster. Cut the laths by using a hacksaw blade backwards. This creates a pull instead of push on the lath. Saw up to the top on both sides. Insert a screwdriver in the saw slot and twist, breaking this half piece off. Use a pocketknife to trim if necessary. Do the same with the bottom lath. Now you have room to stick your forefinger in and hook the center lath. This is to prevent the lath from moving when sawing. Sawing might break the lath away from the plaster "key" (the part of the plaster that oozes through between the laths and down the back, forming a little hook).

If the lath breaks away from the plaster, the plaster is weakened and might crack or break away. Hence, you must support the lath while sawing. Make sure you have the blade backwards. Most metal wall boxes have reversible mounting brackets. You want to use this method to make the box flush with the wall surface.

Chip out the plaster so that the reversed brackets contact the lath itself. Half-inch #6 screws are all that is needed and the box is more secure. Patch the areas over the brackets and around the box if needed. This is required by the NEC.

Removing Existing Box

If a switch or receptacle box must be removed from its location in a finished wall to make additions or changes, it can be done without much trouble. Plastic wall boxes are fastened to a stud by two nails driven through nail guides cast into the top and bottom outside of the box. These nails can be cut off using a hacksaw blade (the blade only, not frame). See Fig. 9-21.

Insert the blade near the top and bottom corners of the box opening in the drywall (not inside the box). Sometimes a screwdriver is needed to make a hole to start the blade. Saw at a position just above and below the corner of the box, so angle the blade in that direction. You will feel and hear the blade contacting the nail. After both nails have been cut, run a pocketknife around the box and you will be able to remove it easily.

To replace the box, drill two holes—³⁄₁₆ of

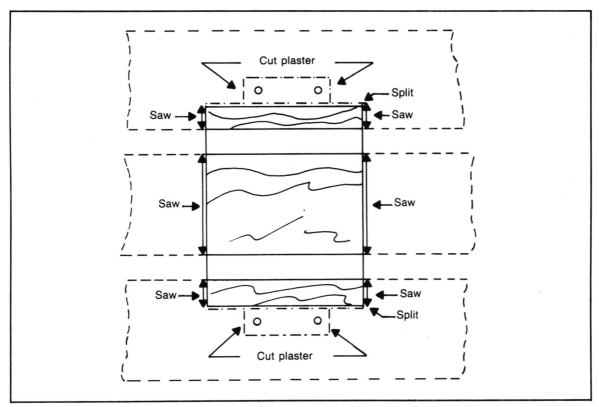

Fig. 9-20. Cutting the wall box opening in lath and plaster.

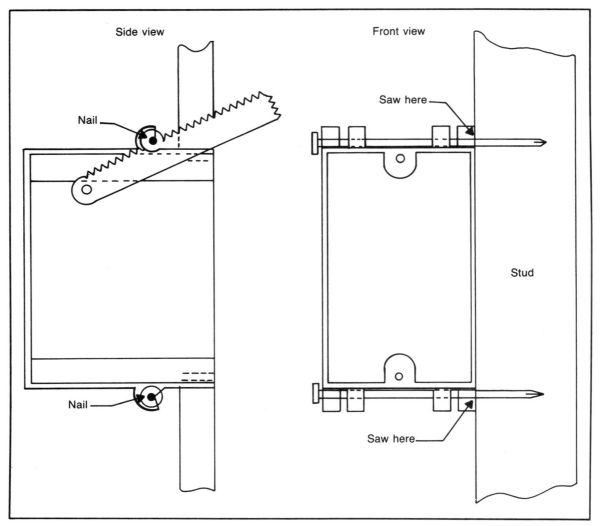

Fig. 9-21. Method of removing plastic wall box.

an inch in diameter—about an inch inside the stud side and about 2 inches apart. Use #8 roundhead screws to fasten the box to the stud. Try a screwdriver angled into the box where the screws would be before drilling the holes to make sure you are able to tighten the screws. Avoid any closeness of the device in the box to the screws.

Finding the Center of a Ceiling

Assume that the room you want to find the center for is 12 × 14 feet. Set a stepladder near the room center. Find a piece of 1 × 2 wood about 6 or 7 feet long to use as a measuring device. Go up the ladder so that you can hold the stick with one end at the wall and ceiling juncture near the center of that wall. Mark the

100

other end (that is over your head) with a pencil mark on the ceiling. Now put the stick against the opposite wall-to-ceiling juncture and make another mark on the ceiling at its end. You now have two marks about one or two feet apart. Whether the stick was more or less than half the span does not matter. With your rule, measure that span and divide the figure in half. Half of 11½ inches is 5¾ inches; half of 19 inches is 9½ inches. Be sure to designate this mark as the center of that span. Circle the mark.

Do the same with the other two walls to find the center in that direction also. Make your marks near the first circled center mark. This second center mark may have to be moved slightly one way or the other to coincide with the other mark. This method can also be used with new work. The first marking should be the bottom of a joist, the same as on a finished ceiling. Determine which joist space is nearest to the room center. Locate the room center in the other direction on this joist.

Temporarily tack up a sliding bar box support at the approximate center in the crosswise direction to the joists. Now by a skillful handling of the stick (1 × 2) and your rule, find the center in this direction. Figures 9-22 and 9-23 should make this method clearer. This whole layout takes no more than five minutes to do.

Because off-center fixtures are an eyesore, it is important to locate them accurately. If the center falls on a joist, buy a half-inch-deep ceiling box. Any ceiling finish that is thinner than one-half inch will make it necessary to notch the joist slightly. The box must be flush with the finished ceiling. Be sure to allow knockouts to be clear of the joist for insertion of necessary cables into the box. For old work conditions where the box comes between joists, use either a spring-out U-support or the bar hanger. If the fixture is heavy, it is safer to use the bar hanger. For heavy fixtures, the NEC makes it mandatory to support them from the building framing members.

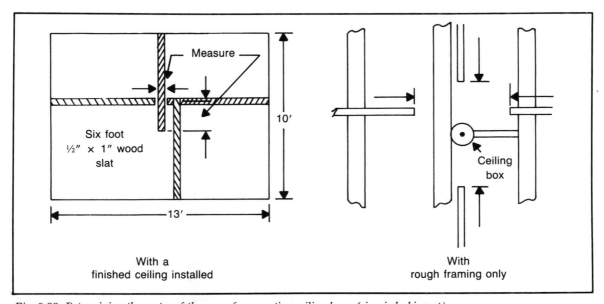

Fig. 9-22. Determining the center of the room for mounting ceiling boxes (view is looking up).

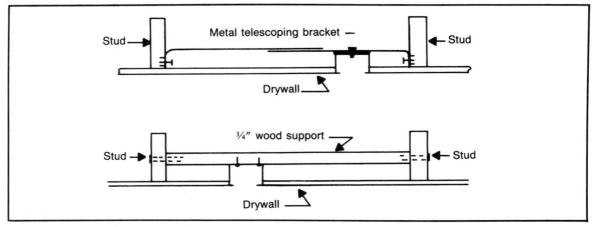

Fig. 9-23. Cutaway view of methods for mounting ceiling boxes.

Mounting Boxes on Masonry Walls

Boxes mounted on masonry walls usually are found in basements of homes or in commercial and industrial buildings. Such boxes must have conduit brought to them from overhead (in home basements), or they must form part of a complete raceway (conduit) system in a commercial building. Anchoring methods have many forms such as plastic inserts, fiber inserts, toggle bolts (for hollow block walls), and hardened-steel drive pins. Drive pins are driven either by a manual pin holder and hit with a hammer or by a *powder-actuated* special tool (they are actually shot into the concrete).

Boxes used on masonry walls are usually the "handy box" type having rounded corners. There are two holes in the back of the box in a staggered arrangement for stability in case the wall is not smooth. Hold the box plumb and mark the hole locations through each hole with a Listo pencil. The Listo pencil is similar to a mechanical pencil except the marking point is ⅛-inch in diameter and is soft like a crayon. They are very good for rough work. Buy one in an office supply store (they cost about $1, and refills cost about 50¢ for 5 leads).

Before drilling the holes, mark a large cross with its lines intersecting on the mark you made through the box holes. For drilling, use a ³⁄₁₆-inch carbide drill and drill a little deeper than the insert. Plastic inserts work fine for this lightweight box. A popular brand made by Rawl is Bantam Plugs, size #8-10 × ⅞-inch. A ³⁄₁₆-inch carbide drill is used. Use #10 sheet-metal screws (¾-inch long) for fasteners. You will be bringing ½-inch or ¾-inch conduit down the wall and into the box, usually from the basement ceiling.

To have the conduit enter the box properly, it must have an offset at the end where it enters the box. An alternative is to buy a conduit offset connector that makes forming an offset unnecessary. Secure the conduit with *one-hole straps*. These are a little more expensive, but they eliminate drilling two holes for each strap.

ANCHORING DEVICES AND FASTENING METHODS

Devices, boxes, cabinets, cables, and wires need some means of being secured to a supporting member. Anchoring devices described in Table 9-1 include many items for many differing support conditions.

Drive pins can be used with a drive-pin holder and a small sledge or used with a powder-actuated (explosive charge in a .22-caliber shell) gun-type tool. These pins are hard to use. Some types of concrete in certain parts of the country are almost impossible to drive pins into. In my area, the concrete has extra hard stones that divert the pin, bending it and causing it to chip out a piece of the concrete. On one job, I tried five times to shoot a pin into the concrete, losing both the pin and the powder charge each time. I ended up using a lead anchor and machine screw. One pin shot costs about 40¢. Two dollars plus my time reloading, holding the box, and shooting was wasted. I had the lead anchor in the wall and the box mounted in about two minutes.

Lead anchors might be prohibited by local fire codes because in a fire, the lead will melt and release whatever it was supporting. In such regions, steel anchors must be used. These work the same way as the lead anchors. Local codes might require steel only in commercial buildings.

Lead anchors, steel anchors, and hollow-wall anchors all use machine screws that are not supplied with the anchors, with the exception of hollow wall fasteners. All these anchors are available in various sizes to suit the job at hand. Lead anchors are commonly known as A & J's because they were developed by the Ackerman-Johnson Company. Lead anchors need a *setter* to pound them into place. One is furnished with a full box. Because the setter costs about 75¢, a satisfactory substitute is a ¼×4-inch-long pipe nipple of steel, not brass. If you don't have a carbide drill, a substitute is a ½-inch star drill

Table 9-1. Fastening Devices and Their Uses

Item	Uses
Wood screws	Mounting cabinets and boxes on wood or composition surfaces.
Sheet-metal screws	Mounting cabinets and boxes on wood sheet metal or composition surfaces.
Concrete anchors	Mounting cabinets on concrete surfaces.
Drive pins	Mounting cabinets and boxes on concrete surfaces.
Toggle bolts	Mounting cabinets and boxes on concrete block walls and drywall.
Plastic plugs and Fiber plugs	Mounting lightweight boxes on concrete or drywall.
Lead Anchors and Steel Anchors	Mounting heavy cabinets on concrete.
Wood plugs	Prohibited in electrical work.
Hollow wall anchors	Mounting cabinets on hollow walls such as concrete blocks drywall and paneling.
Beam clamps	Mounting equipment on steel beams.

and a heavy hammer. In an emergency, you can use a ½-inch-wide cold chisel. Both the star drill and the cold chisel must be rotated while pounding on them.

For mounting boxes and devices on sheet metal, use sheet-metal screws. These screws are very hard and tend to chew up screwdriver blades, so use a screwdriver that fits the screw slot snugly to help prevent blade wear. Phillips-type screws are unsatisfactory because the screwdriver tends to slip out of the cross slot when trying to start such a screw when using great pressure. For standard slotted screws in the "pan head" style, boxes of 100 are less expensive by the piece than those in bubble packs of 10 or so. If you can use that many, buy the larger quantity. These screws have many uses in other work. Plastic and fiber plugs require sheet-metal screws.

When drilling for plastic or fiber plugs and you have made an error in placement of the hole and have inserted the plug, it can be removed. Start a screw in the plug about halfway; grip the screw with electrician's side cutters as if you were going to cut the screw head off. Now lift the handles of the cutters and most of the plug will come out with the screw. The hole can now be patched. Plan carefully and you won't miss often.

Fiber plugs (Rawl Plugs is one brand) have a fiber shell with a lead lining. The plug, screw, and drill should be all the same size. As the screw is driven in, the tightness is very considerable. If the conditions are all not right, the plug might, in plaster or drywall, start turning. The screw then cannot enter and expand to make a tight fit. Some plastic plugs have fins on the sides and a flange on the top to help with the installation and prevent turning.

Boxes for masonry walls are usually handy box style with rounded corners. There are two holes in a staggered arrangement in the box back. This will make the box more stable when it is mounted. Concrete walls are not always level. Hold the box plumb and level and mark the location of the two holes. Using a ³⁄₁₆-inch masonry drill, drill a 1-inch-deep hole at both marks. Masonry drills tend to wander from the mark. Make a cross over the original mark using a Listo china marker pencil (available at office supply stores). Make the legs of the cross 1 inch long. If the original mark is gone, the intersection of the cross lines locate it again. Try to guide the drill back to center by angling it slightly (not too much or it will skid away entirely). The Listo pencil is fine for marking on rough surfaces such as concrete and for marking pipe and conduit.

Toggle bolts fit in holes in hollow walls such as drywall and concrete blocks. The wings of the toggle bolt are larger than the shank of the bolt so a larger hole must be made to insert them. Thus the bolt is a sloppy fit in the hole, so it must be tightened very tightly to prevent movement of the object being mounted on the wall. The toggle bolt assembly cannot be recovered if the item is removed, because the toggle wings fall down inside the wall.

Hollow-wall fasteners are sometimes better than toggles because they fit tightly in the hole and allow no lateral movement of the cabinet or box, and cabinets, shelves, or other items can be removed from the wall and later replaced when decorating. The hollow-wall fastener has a hollow cage arrangement consisting of four metal strips slightly bent out at the center and connected at the top and bottom. The top has a flange wider than the cage and the bottom has a round nut with internal threads. The hole is drilled and the anchor is tapped into the hole. Then the screw is inserted and tightened until the four strips behind the wall start to spread outward. Further tightening pulls the nut up tight on the back side of the wall, thus forming a solid anchor, similar in appearance to a toggle bolt viewed from the rear.

When using a toggle, sometimes a part is left off the bolt and the toggle is tightened to the wall. To remove the toggle results in the loss of the wings inside the wall. Be sure all parts are in place before inserting the toggle.

Bending Thinwall Conduit (EMT)

The bending of Thinwall is somewhat of an art, because it takes forethought, accurate measurement, and an even temper. Once a bend has been made, there is no rebending or correction of the bend possible. Slight adjustments might be possible, but beyond that, the piece is best set aside for possible use in another location if done incorrectly.

If you plan to do much Thinwall work, it pays to get an instruction booklet for the make of bender you will be using. Benders cost about $15 or they can be rented in many areas. The rental agency might have a book or they will show you how to use the bender. The two basic operations are making the 45-degree or 90-degree bend and making the offset and saddle bends.

The *offset* bend is made to raise the end of the Thinwall off the wall so that it can line up with the box or cabinet knockout. For the offset, there are offset box connectors made that eliminate the need for a job-made offset. These connectors are for the standard offset and will not work if the distance is more than about ⅜ of an inch. The job-made offset looks the best, but it takes time. The *saddle* bend is a neatly formed "hump" to allow the Thinwall to be run over another pipe or other type of obstruction.

To make a 90-degree stub bend 12 inches high, make a mark on the Thinwall 7 inches from the end. The bender takes up the other 5 inches. This is for ½-inch Thinwall benders. The arrow on the bender is then set on this mark with the bender hook around the Thinwall itself. Put your foot on the bender foot pad. Use either foot depending on your convenience and whether you are right-handed or left-handed.

Now pull down on the bender handle until the handle is at a 45-degree angle. This completes the bend and the stub will be vertical. The second bend on the same piece is more tricky because the two bends must be in correct relation to each other. If the second bend is not made in the right direction, the piece will not fit. This can be corrected by cutting the Thinwall between the two bends and using a coupling. The second bend can then be pointed in the right direction (see Chapter 3).

To make an offset bend to enter a cabinet or box, hook the bender near the Thinwall end, but not *on* the end because the hook will distort the end itself. To avoid this, attach a box connector on the end tightly; then the bender hook can be hooked on just behind this connector. As before, set the bender on and bend. The idea is to make a 45-degree bend, almost. Now "upend" the bender with the handle end on the floor and bender in the air. Put the Thinwall back in the bender so that the opposite direction bend can be made. (See Fig. 3-11).

Another method of making the second bend for an offset is to find a raised platform, such as a porch or steps, so that the conduit end can be extended over the edge. Then make the second bend on the floor rather than in the air.

Sometimes it is advantageous to slip a larger pipe over the conduit being bent so that the bend will not form away from the bender itself rather than in the bender groove. This pipe

105

is called a *slip pipe*. A *saddle* is two offsets back to back, one to raise the Thinwall run over an obstruction (pipe, beam, etc.) and another offset to return the run to the previous level. Thinwall bending takes some skill and practice to accomplish the proper bends. Once you get the hang of it, you will have little trouble. Expect to spoil one or two lengths practicing, but this costs much less than hiring an electrician.

When bending, use a slow, even pull on the bender handle. Do not jerk on the handle because you could kink the conduit. Keep the conduit flat on the floor and your free foot on the bender foot pad. If the conduit raises up from the floor, it might kink just behind the bender itself instead of bending in the bender groove. Kinks cannot be repaired, because the inside wall is misshapen and it will damage the wires when they are pulled through. An inspector will turn down damaged conduit. Slow and easy does it. You will have failures at first, so be patient.

Bending Rigid Conduit

Rigid conduit is entirely different from Thinwall conduit. A special bender called a *hickey* is used. Because the hickey is much shorter than a Thinwall bender, the bend must be made in short bends, or *bites*, of the hickey on the conduit rather than one full sweep as in Thinwall. The use of rigid conduit is discouraged because of the additional work and tools needed for its installation. To use rigid conduit, you would need a pipe vise, threading dies, cutting oil and an oil can, and a pipe reamer. The homeowner generally has no need to install rigid conduit unless special conditions cause the local inspector to require it.

Bending instructions are much the same as for Thinwall except for allowances for threads going through the walls of boxes and cabinets

(to take the place of box connectors on Thinwall). Larger sweeps can be made with the hickey than with Thinwall because there is no set radius for bends. Both conduits restrict the number of bends in one run between pull boxes to four 90-degree bends or the equivalent. The number of wires allowed in each size conduit is restricted by NEC Tables 3A, 3B, and 3C. Three tables are necessary to allow for the different physical diameters of the same size wire due to different types and thicknesses of insulation on them.

Rigid conduit is threaded similar to plumbing pipe. Plumbing threads have a certain taper to the threads, while electrical rigid conduit threads have less taper than the plumbing pipe threads.

Different styles of pipe dies are available. Examples are:

▶ Two handled (die in the center and a handle on each side).

▶ Ratchet type (one handle worked like an automotive ratchet wrench). This is ratcheted on to cut the thread by pushing down on the handle and lifting up and then down again. The direction dog is reversed and the die and handle are spun off backwards, counterclockwise.

▶ This type is ratcheted on and after the thread is cut the die is opened and the whole tool is then lifted off the conduit.

▶ The last is a power-operated pipe-threading machine that costs about 3 thousand dollars.

In all types, the die is run on the conduit until one or two threads show beyond the die edge or face. Cutting oil must be used or the die becomes overheated and loses its edge, and it will need to be replaced. This is extremely important in the case of the power threading machine. In certain cases, underground wiring needs to be run in rigid conduit, such as for swimming pools and similar installations.

Intermediate Metal Conduit

This type of conduit was developed as an excellent substitute for heavy and heavy-walled rigid conduit. It has been approved for any and all uses for which rigid conduit is approved. It is easy to handle and work. Although standard threaded fittings are used, the metal wall is thinner, thus saving in cost as well as saving of steel. The conduit is galvanized and can be buried in concrete. Threadless fittings can be used that are similar to those used with Thinwall tubing. This is a very satisfactory conduit, nevertheless, a homeowner might never have occasion to use it.

Armored Cable (BX)

Armored cable, commonly known as BX, is similar to conduit except that it is flexible. It is approved for residential uses in nearly all areas. Installation in damp or wet locations is prohibited unless type ACL (lead covering inside outer armor) is used. The steel covering consists of a spiral steel strip with interlocking edges wrapped around to form a continuous covering for the wires inside. The armored covering itself, with no wires inside, is known as greenfield. BX is sold in sizes from #14 to #1, having two and three wires in the cable.

Where BX is run through studs and floor and ceiling joists, holes must be bored through the members in the center of the face to prevent nails from being driven through the cables. When cable is run in notches cut in the edge of the framing, metal plates must be put over the cable and notch. Commercial faceplates are made with points to be driven in the wood to

hold the plate in place. These plates are mandatory for nonmetallic cable and optional for BX.

When BX is cut and the armored covering is removed, a fiber bushing must be inserted between the cut edge of the armor and the insulated wires inside. If the bushing is not put in place, the rough edge of the armor will damage the insulation. Because the bushing is red, it can easily be seen through holes in the BX connectors and BX-style clamps (different than Romex clamps) in the boxes and cabinets. Always insert the bushing.

To cut BX, use a hacksaw blade with 32 teeth per inch. Coarse tooth blades will cut too fast and cut into the wires inside. Figures 9-24 through 9-26 show procedures for cutting BX cable. A BX cable to be cut will either come from a cable already installed and ready to be cut off to the correct length, or from a roll being used on the job.

To cut off an installed length from a coil, be sure the part already in place is secured. Make a mark with a Listo pencil for the length. Grasp the BX 6 to 8 inches beyond the mark on coil side with your left hand. Pull the cable tight. Then cock your hand up at the wrist back toward your body to put a "hump" in the cable at the pencil mark. Carefully start sawing at the mark. Saw at right angles to the spiral but *not* at right angles to the run of the cable. Saw the highest part of the spiral almost through, leaving just the lower parts not sawed. Take the cable in both hands and bend it back and forth at the cut until the strip snaps free. By twisting the two cut ends counterclockwise, a separation is formed and the wires can either be cut with diagonal cutters or sawed in two. From the coil on the floor, you must put your foot on the cable at the floor to be able to make it tight, then proceed as above.

To strip the armor in preparation for insert-

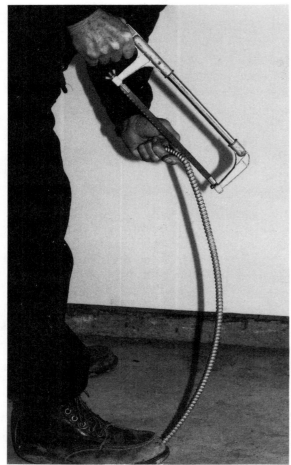

Fig. 9-24. Method of holding BX to cut the steel covering, using a hacksaw. The cable is being held by the foot and is curved at the cutting point. If the cable is fastened at the far end, the foot pressure is not needed.

ing it into the box or cabinet, mark 8 inches back from the end. Now very carefully saw almost through the armor *only*. Bend the cut back and forth until the armor breaks. Do not bend too sharply because that might cut into the insulation at this point. If the insulation is damaged here, you must cut the damaged part off and start again. The internal bonding strip should be bent back over the armor before putting on the box connector.

The original BX did not have the internal bonding strip, as it was not considered necessary. It was then decided to eliminate any high resistance due to poor contact between individual turns of the metal covering by adding the bonding strip.

Tighten the connector (binding all parts together). The best way to tighten the connector to the box is to run the locknut up tight on the threads, by hand, then using Channellocks tighten the outside part of the connector. Adjust the locknut so the connector setscrew ends up on the topside, and insert the BX with its bonding strip bent back. Finally, tighten the connector setscrew. There are cutters for stripping the armor from the BX in preparation for inserting into the box or cabinet, but they are expensive. The hacksaw has worked fine for many years; just be careful when using it.

Nonmetallic Sheathed Cable (Romex)

Nonmetallic sheathed cable is commonly known as Romex. Romex was developed by Rome Cable Co., Rome, New York. It was originally called Rome-X. Both Romex and BX have many characteristics in common. BX was the forerunner and Romex was probably developed from BX. Both cables have two or three

Fig. 9-25. Bending BX armor to snap covering at saw cut.

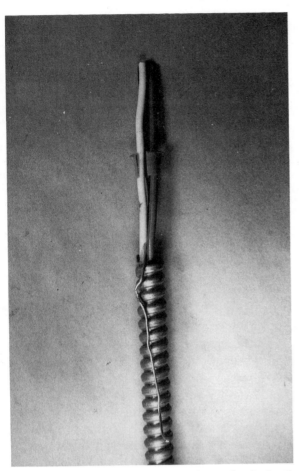

Fig. 9-26. BX Armor has been removed from end and red anti-short (insulating) bushing is in place as shown in Fig. 7-9. Bonding strip has been bent back and will be clamped to the BX metal covering by the BX clamp. The bonding strip is similar to the bare Romex ground wire.

insulated wires as required for use in various circuits. The two-wire type has one black and one white wire.

The three-wire cable has one black wire, one red wire and one white wire. BX has, in addition, the flat aluminum continuity strip for grounding of the outside armor to provide a continuous equipment ground system. As BX has the aluminum strip, so Romex has the copper

ground wire. Never buy Romex that does not say "with ground." This type of cable is illegal in most areas.

The uses for both cables are the same: residential and light commercial. Some local codes prohibit Romex (Chicago is one).

There are tools for stripping the Romex sheath. At one time, I was using a seamstress' seam ripper, but the outer covering on NMC type cable is thicker and tougher, and the seam ripper was not strong enough to do the job. One cable stripper on the market sells for under $2 and works well. I had to sharpen the cutter blade on mine before it would work properly. I now have, in addition, a utility knife with a hook blade. This hooked blade is dug into the center of the flat of the cable and pulled along, cutting the covering but not the insulation. Be careful, and practice some before using it.

Strip about 8 inches of covering from Romex when you are ready to insert it into a box or cabinet. Make sure ¼ inch of covering shows inside the box clamp. Cable must be supported (stapled) every 4½ feet and within 12 inches of the box or cabinet. If the plastic box has no clamps, Romex must be stapled within 8 inches of the box. There are now Romex staples consisting of a plastic top bar with a nail in each end. These are better than the all-metal staple that is hard to drive straight and is liable to damage the cable covering. The new plastic-top staple will give some and prevent cable damage. Each nail can be driven separately. With either staple, do not drive too tight and damage the cover.

Romex must be protected from damage. If it is carried in notches in framing members, it must be protected by a metal plate, 1/16-inch thick, fastened over the cable and notch to protect the cable from nails. If Romex is threaded through holes bored in framing members, the holes must be at least 1¼ inches from the framing member front edge. In this case, a metal sleeve $\frac{1}{16}$-inch thick (use $\frac{1}{2}$-inch Thinwall about 2 inches long) driven into the bored hole. Ream both ends of the sleeve.

Running cable in "old work" is easier and faster when two people work together. When two people are not available, drop a cord with a weight (mouse) on it from a switch or receptacle box to the bottom of the hollow partition. A bent coat hanger wire, when bent properly, will hook the cord and pull it down through the hole in the base plate of the hollow partition (see Fig. 9-1). Because old work is concealed and cannot be stapled, boxes should have cable clamps. The boxes with diagonal back corners should be used if available. Their shape makes it easier to insert the box in the wall with two or three cables clamped in it. Be sure the cable sheath shows inside the box clamp. This is an NEC requirement.

When you are ready to attach the wires to the device, allow 8 inches from the point where the wires enter the box for making the connections. This gives enough slack to allow the wires to be bent, S-curve shape, back into the box after attaching the wires to the device. You can form the wires into loops with needle-nose or side-cutter pliers. The loops must go around the screws clockwise to prevent opening the loop when the screw is tightened. If the device has push-in connections, strip $\frac{5}{8}$ of an inch. If it has screw connections, strip $\frac{3}{4}$ of an inch.

Wire Identification

When it is necessary to disconnect wiring—such as on motors, heating and cooling equipment, and loose wires in junction boxes—each wire must be marked to identify it for reconnection. Various methods of identification are used; examples are paper covered with clear tape, tags, and bending each wire end into a distinctive shape. Each of these methods can sometimes result in the loss of the identifying marker.

Permanent marking materials are available that consist of printed numbers, letters and words with pressure-sensitive backing. The material is a clothlike substance packaged in different forms. Pocket booklets have numbers 1 to 25 or 1 to 50. Other cards consist of individual numbers, each card having only one number such as 3 or 47. Electrical supply houses sell these numbers and they will usually sell to individuals. Some electronic stores might also have these numbers.

Whether you make your own or buy your identification tags, be sure to identify wires for reconnection. If you don't, you are in trouble. Even two wires, if they are part of control circuits, should be identified. The stick-on numbers really stay in place on the insulation of the wire. Just make sure the wire is not greasy. Even though the stick-on number is overlapped on itself, it could slide off from a greasy surface.

Chapter 10

Permits and Inspections

As with any alterations or additions, a permit is needed and required to legally do this work. Permits are obtained from the governing authority having jurisdiction in the area. Aside from the legal aspects of obtaining a permit, there is the satisfaction of having the work inspected and approved.

PROCEDURE FOR OBTAINING PERMITS

To apply for a permit to do electrical wiring, visit the building inspection department of the local governing body. This will usually be city hall or the local township offices. Most, but not all, governing authorities issue what is known as a homeowner's electrical wiring permit. There might be variations in the title of the permit, but the purpose is the same. This permit allows you to do electrical wiring only in your own residence but excludes rental property that you might own. Ordinances are very strict about this.

Some areas do not permit persons who are not licensed electricians to do electrical, plumbing, or heating and air conditioning work. Hawaii is one state that has this restriction statewide. In such areas, it might be possible to take an examination and thus obtain your own license (if you pass the exam). You can then get a regular permit to do the work. Permits are necessary for two reasons:

▶ To protect your insurance coverage. If the house burns down and the insurance company finds that the wiring was done without the proper permits, it could void the insurance coverage and you will not be covered for the fire damage.

▶ The inspector must inspect the work at least once (usually twice), and will approve or not approve the work. If the inspector does not approve the work, he usually tells you what is wrong and how to correct the work.

When you apply for a homeowner's electrical permit, you must swear that you will do all the work yourself and not contract out any of it. You must do the work yourself. No officials will complain, however, if you have one of your children or your spouse give you a helping hand now and then.

If the wiring is extensive or in a new building, you will be required to provide an outline and perhaps rough drawings of the proposed wiring layout to the building department. This doesn't need to be a detailed wiring diagram, but just a layout showing receptacles, switches, range and dryer receptacles, and the service entrance location and capacity. The inspector usually suggests proper wire sizes and types and will clue you in on any variations of the local code from the National Electrical Code.

Forms (Figs. 10-1, 10-2 and 10-3) from two cities in different states show types of permits issued by municipalities for homeowners to do their own electrical wiring. The permit form (actually called the "homeowner affidavit") from the city of Troy, Michigan, outlines the rules

CITY OF TROY

500 W. Big Beaver   X689-4900X   524-3344

HOME OWNER AFFIDAVIT

LOCATION_____DATE_____

As the bona fide owner of the above mentioned property which is a single residence, and which is, or will be on completion my place of residence and no part of which is used for rental or commercial purposes nor is now contemplated for such purpose, I hereby make application for an owner's permit to install_____ as listed on the permit application.

I certify that I am familiar with the provisions of the applicable Ordinance and the rules governing the type of installation which is contemplated at the above mentioned location and hereby agree to make the installation in conformance with the Ordinance.

In making this application, I realize I am assuming the responsibility of a licensed contractor for the installation of the work mentioned in the permit application and for putting the equipment in operation. I further agree that I shall neither hire any other person for the purpose of installing any portion of the_____or related equipment at the above premises, nor sub-contract to any other person, firm or corporation the installation of any portion of the above equipment.

I agree to notify the Inspection Department within seventy-two (72) hours after the installation is completed and is ready for service so that the Department may make its required inspection. I further agree to keep all parts of the installation exposed until the installation is accepted as being in compliance with Ordinance requirements.

I further agree to correct within two weeks time any violations on the work installed and to provide access to the premises between the hours of 8 a.m. and 5 p.m., Monday through Friday for the necessary inspection or inspections. Failure to correct violations or to provide access will subject the permit to cancellation in which case a licensed contractor must be employed to complete the work.

APPLICATION: □ ACCEPTABLE   □ NOT ACCEPTABLE _____

White copy to applicant.
Yellow copy retained in office.   Department Representative

If not acceptable, list reason_____

Subscribed and sworn to before me this

_____day of_____, 19___.   _____
                                                Owner

                                         Present Address

Notary Public,_____County, Michigan.

My commission expires_____.   Telephone Number

Fig. 10-1. Homeowner affidavit, city of Troy, Michigan. This permit can be used for any work done by the homeowner needing a permit and inspection of finished work. (Courtesy City of Troy).

and regulations and is restricted to a single family residence. I have underlined the statement of the owner agreeing to do all the work and not to sub-contract any work listed on the permit to another person.

The second permit—from the city of Zion, Illinois—assumes that the applicant is a licensed electrician. Conversation with the chief inspector confirms that a permit can be obtained by a homeowner. The permit contains a statement that the installation will conform to all applicable codes, including the National Electrical Code. If not mentioned, the NEC is assumed to be the latest and current edition. Note that there is a list of fees for various inspection permits. See Fig. 10-2.

The inspector will be required to make at least two and perhaps more inspections. The first inspection is made before the wall finish (such as drywall) is applied. This is so that the roughed-in wiring can be checked for correctness and conformity to NEC requirements. The inspector will check on support of the cable and proper mounting of switch/receptacle/junction boxes. Protection for cable where it is run through 2 × 4 studs or 2 × 6 and 2 × 8 joists to make sure no nails can penetrate the cable (especially Romex) will be inspected.

The arrangement of the service-entrance/distribution equipment and the proper mounting and anchoring of the cabinets and the securing of the entrance cable along the outside of the building is checked. The inspector will check to make sure the entrance head at the top of the cable is above the point of attachment of the service drop to the building wall, plus many other items. I have only named the major ones. Call for this first inspection well in advance, because the inspector could be busy and might not be able to come when you request. If the inspector makes an unnecessary trip, you could

be charged additional fees.

The second inspection covers receptacles, switches, circuit breakers (the right size?) and all distribution and supply wiring (also for correct size). If you have read and followed the NEC, you should have no trouble getting approval. If your work is not approved, listen to the inspector for suggestions as to how to correct your installation. It would help, on the first inspection, to ask if there is anything that could be corrected or made better. Most inspectors advise of changes during the rough-in (first) inspection, but it would not hurt to ask, "Is there any way the job can be improved over the way it is now?"

A page from installation instructions of the Hawaiian Electric Light Company, Inc., for the installation of underground cable from their lines to the customer's meter, gives their requirements for underground cable installation.

THE INSPECTOR

The local inspector is usually a former electrical contractor who has retired from contracting and has been appointed to the position of electrical inspector. Many inspectors have had extensive education both in the practical application and the theoretical knowledge of electricity. One inspector in my area is on the board of the National Electrical Code. Almost all inspectors have worked with tools and have intimate knowledge of all phases of electrical construction and design. The position of inspector is usually political and appointive.

The inspector will be helpful to those who do not have extensive knowledge of wiring practices. A few words might, in the long run, save you time and money as the inspector knows the best and most economical methods of wiring. Listen and talk with inspectors every chance you

APPLICATION FOR ELECTRICAL INSTALLATION PERMIT

Building Permit No. _____ Electrical Application No. _____

Date of Inspection _____ 19 _____ Application checked by _____

CITY OF ZION

Phone: 872-4546   Date _____ 19 _____

APPLICATION is hereby made for the approval of the plans and specifications hereby submitted, and made a part hereof for the construction or alteration, as appears in specification sheet attached hereto, of electrical installations in the building on premises described below:

APPLICANT agrees to comply with all provisions of the Municipal Code of the City of Zion with reference to electrical installations and such other laws relating to erection and alteration of buildings in effect at this date.

APPLICANT agrees to post the Electrical permit on the premises before work is started and notify the Electrical Inspection Department when ready for inspections, as provided by the electrical ordinance.

Location: Street and Number _____ Lot _____ Block _____ Sec. _____

Contractor _____ Address _____

Owner _____ Address _____

For Initial Installation of Electrical Service: Fees   PAID

(a)   Single Family Residence or Bungalows (New construction):
      Without electric heat _____ $ 7.50
      With electric heat _____   10.50   _____
      (plus 15¢ per kilowatt over 600 watts)

(b)   Multi-Family or Apartments:
      First Apartment (new construction)
      Without electric heat _____   7.50
      With electric heat _____   10.50   _____
      (plus 15¢ per kilowatt over 600 watts)
      Each additional apartment:
      Without electric heat _____   6.00
      With electric heat _____   9.00   _____
      (plus 15¢ per kilowatt over 600 watts)

(c)   Commercial or Industrial:
      The fee for commercial or industrial wiring
      shall be $5.00 per thousand valuation.
      Minimum charge to be _____   7.50   _____

(d)   Electrical Neon Signs _____   3.00   _____
(e)   Swimming Pool Fee _____   12.00   _____
      (subject to Article 680 of the National Electrical Code)

Revisions - Additions:
(a)   Incidental Units:
      Air Conditioning Units (1 ton or more), Water Heaters, etc.
      First Unit _____ $ 3.00
      Each additional unit _____   1.50   _____

(b)   Up to five outlets, not included in (a) or (b) as shown in Section 1.
      Minimum charge _____ $ 3.00   _____

(c)   Entrance services _____   3.00   _____
(d)   Electric space heating:
      The basic fee will be $3.00 plus 15¢ per kilowatt over 600 watts.
                                                                     _____

Plan filed _____ 19 _____   TOTAL _____

I hereby certify that I have read the Electrical Sections of the Municipal Code and know the contents thereof.

Applicant _____   Address _____
          Licensed Electrician

Applicant's Phone No. _____   License No. _____

Fig. 10-2. Application for Electrical Installation permit, City of Zion, IL (Courtesy City of Zion).

Sec. 7-48. Regulations of electrical utility adopted.

The rules and regulations regarding the installatino, alteration and use of electrical equipment as last adopted by the electricity supply company supplying the city as last published, and filed with the Illinois Commerce Commission and as approved by the electrical commission of the city are hereby adopted. (Code 1953, § 21-501; Ord. No. 64-0-68, § 1, 10-6-64)

Sec. 7-49. Filing, incorporation of code and regulations.

A copy of the code and rules and regulations adopted by this article are on file in the office of the chief electrical inspector and the provisionsof said code and said rules and regulations are hereby made a part of this Code. (Code 1953, § 21-501; Ord. No. 64-0-68, § 1, 10-6-64)

Sec. 7-50. Exceptions, modifications in code and regulations.

The following exceptions and modifications not clearly indicated in the National Electrical Code or in the rules and regulations of the electricity supply company are found necessary to meet conditions, and shall prevail in case of conflict thereof with any other provisions of this chapter or other ordinances of the city.

(1) Section 348-1 of the National Electrical Code is amended to read as follows: "*1. All electrical wiring shall be in electrical metallic tubing* as specified in article 348 of the latest edition of the National Electrical Code as amended, *or rigid metal conduit* as specified in article 346 of the National Electrical Code, except that other types of conduit or wiring protection may be permitted by the electrical inspection department for portions of the wiring where use of electrical metallic tubing or rigid conduit is impractical. 2. Electrical metallic tubing may be used for both exposed and concaled work, except as follows: Electrical metallic tubing protected from corrosion solely by enamel shall not be used. Electrical metallic tubing shall not be used-where, during installation or afterward, it will be subject to severe physical damage, and shall not be used in or under concrete in slab or grade construction. Only rigid steel conduit is to be used under concrete in slab or grade construction and not rigid aluminum conduit.

"In permitted closed wall construction a minimum one-half inch Greenfield flexible metallic conduit with appropriate size copper wire pulled and all metal connectors and boxes may be installed. Ninety (90) degree els are not allowed. Thin wall or rigid conduit shall be installed where such work is open and accessible.

"*In all existing buildings additional wiring may be installed in rigid or thin wall conduit, B.X. flexible metallic conduit, approved flexible metallic conduit or approved metal moulding* except that where exposed to moisture or weather rigid conduit or thin wall conduit shall be used. Where flexible metallic conduit or B.X. is used, it shall terminate in a box in the basement to be located not more than three (3) feet from where the conduit or B.X. enters the basement. All electrical materials and fixtures are to carry the underwriter's label."

(2) *All service shall be rigid or thin wall metal conduit on the outside of the building and firmly fastened to the building, installed according tot he national code.*

All underground services under streets, alleys, or public ways must be enclosed in rigid metal conduit and must be installed at least thirty (30) inches below the lowest level in said street, alley or public way at the point where said service crosses them.

(3) All wiring for gasoline station pumps, outside lighting and signs installed in rigid steel conduit with conductors of gasoline and oil resistant type wire as approved for such installations and as of article 514 of the latest edition of the National Electrical Code.

(4) *No entrance service shall be smaller than 100 amp capacity. Minimum size wire to be No. 3RH copper.*

Fig. 10-3. Excerpts from the NEC that have been made a part of the City of Zion, IL building code. (Courtesy City of Zion).

get. Be informed of local regulations that differ from the National Electrical Code.

Be sure to follow NEC practices and regulations. The NEC is not difficult to follow. In the NEC handbook for 1987, as in previous editions, there are many explanations and illustrations (neither of which are in the less expensive "text only" edition). It might be advisable and to your advantage to purchase the handbook at $39.50 postpaid.

Use top quality materials and devices in all your work. To match that quality, do work neatly and properly. Almost all electrical wiring materials and devices sold carry the Underwriter Laboratories mark, denoting that the product is suitable for the use intended. Look for the UL label.

Chapter 11

The
National Electrical
Code

The first electrical code was published in 1895 by the National Board of Fire Underwriters. This same Board continued to publish the National Electrical Code until 1962. The code continues to be published every three years, each time being revised and updated as necessary to keep up with new developments in the field of electricity. The National Electrical Code continues to be the most widely adopted code of standard practices in the United States and the world. See Figs. 11-1 through 11-5.

While the code itself is only advisory, it can be adopted by reference and become law by reason of its adoption by a municipality or other governing body. Every governing body establishes a building inspection department for the purpose of inspection to determine the safety and reliability of all building construction. This department is usually divided into many individual sections, and one of them is usually an electrical inspection section that enforces the National Electrical Code and local jurisdictional codes.

LOCAL CODES

Special codes developed by groups of municipalities are enforced by the governing bodies, in addition to the National Electrical Code. In the Detroit area, there are 118 cities and villages that have formed the Reciprocal Electrical Council. This council has published supplementary rules consisting of 13 pages of amendments to the code. These amendments are minor, but they must be considered when doing wiring.

AVAILABILITY
OF THE CODE TEXTS

Copies of the current NEC can be obtained from city electrical inspection departments. Local public utilities might have copies for sale. The

NFPA 70A

Electrical Code
for
One- and Two-Family Dwellings

1987 Edition

Excerpted from the 1987 *National Electrical Code®* NFPA 70 - 1987

Explanation of this Code

This Electrical Code for One- and Two-Family Dwellings (NFPA 70A-1987) covers those wiring methods and materials most commonly encountered in the construction of new one- and two-family dwellings. Other wiring methods, materials and subject matter covered in the 1987 *National Electrical Code* (NFPA 70-1987) are also recognized by this Code. (See preface for further information.)

The development of this Code was first undertaken in 1968 to meet the expressed need for an electrical code applicable only to dwellings as a convenience to those whose interests are so oriented. With the approval of the Correlating Committee of the National Electrical Code Committee, an Ad Hoc Committee was established of those primarily concerned to guide this project to completion. Those asked to serve on the Ad Hoc Committee included representatives of the following organizations: American Insurance Association, Building Officials and Code Administrators International, Inc., Department of Housing and Urban Development, Edison Electric Institute, International Association of Electrical Inspectors, International Brotherhood of Electrical Workers, International Conference of Building Officials, National Association of Home Builders, the National Electrical Contractors Association, and Underwriters Laboratories Inc.

It was decided that the Electrical Code for One- and Two-Family Dwellings should consist of excerpts from the complete current *National Electrical Code* without any modification of intent and with minimum editorial change. Article and section numbers have been retained to permit close correlation.

Following decisions made by the Correlating Committee and by the Technical Subcommittee on format and content, the excerpted and editorially revised material was formally submitted to members of the Technical Subcommittee and the Correlating Committee for letter ballot, to determine if the editorial changes had been achieved without altering the intent of the complete Code.

Fig. 11-1. Explanation of Electrical Code for One- and Two-Family dwellings, NFPA 70A-1987. Reprinted with permission from NFPA 70A-1987, Electrical Code for One- and Two-Family dwellings, Copyright © 1986. National Fire Protection Association, Quincy, MA 02269. This reprinted material is not the complete and official position of the NFPA on the referenced subject, which is represented only by the standard in its entirety.

Fig. 11-2. Table 220-19 (part of Article 220) with Notes 1, 2, 3, and 4. Demand Loads for Household Electric Ranges, etc. excerpted from the 1987 NEC, NFPA 70A-1987. Reprinted with permission from NFPA 70A-1987, Electrical Code for One- and Two-Family Dwellings, Copyright© 1986, National Fire Protection Association, Quincy, MA 02269. This reprinted material is not the complete and official position of the NFPA on the referenced subject, which is represented only by the standard in its entirety.

Table 220-19.   Demand Loads for Household Electric Ranges, Wall-Mounted Ovens, Counter-Mounted Cooking Units, and Other Household Cooking Appliances over 1¾ kW Rating. Column A to be used in all cases except as otherwise permitted in Note 3 below.

NUMBER OF APPLIANCES	Maximum Demand (See Notes)	Demand Factors Percent (See Note 3)	
	COLUMN A (Not over 12 kW Rating)	COLUMN B (Less than 3½ kW Rating)	COLUMN C (3½ kW to 8¾ kW Rating)
1	8 kW	80%	80%
2	11 kW	75%	65%
3	14 kW	70%	55%
4	17 kW	66%	50%
5	20 kW	62%	45%
6	21 kW	59%	43%

Note 1.   Over 12 kW through 27 kW ranges all of same rating. For ranges individually rated more than 12 kW but not more than 27 kW, the maximum demand in Column A shall be increased 5 percent for each additional kW of rating or major fraction thereof by which the rating of individual ranges exceeds 12 kW.

Note 2.   Over 12 kW through 27 kW ranges of unequal ratings. For ranges individually rated more than 12 kW and of different ratings but none exceeding 27 kW, an average value of rating shall be computed by adding together the ratings of all ranges to obtain the total connected load (using 12 kW for any range rated less than 12 kW) and dividing by the total number of ranges; and then the maximum demand in Column A shall be increased 5 percent for each kW or major fraction thereof by which this average value exceeds 12 kW.

Note 3.   Over 1 ¾ kW through 8 ¾ kW. In lieu of the method provided in Column A, it shall be permissible to add the nameplate ratings of all ranges rated more than 1 ¾ kW but not more than 8 ¾ kW and multiply the sum by the demand factors specified in Column B or C for the given number of appliances.

Note 4.   Branch-Circuit Load. It shall be permissible to compute the branch-circuit load for one range in accordance with Table 220-19. The branch-circuit load for one wall-mounted oven or one counter-mounted cooking unit shall be the nameplate rating of the appliance. The branch-circuit load for a counter-mounted cooking unit and not more than two wall-mounted ovens, all supplied from a single branch circuit and located in the same room, shall be computed by adding the nameplate rating of the individual appliances and treating this total as equivalent to one range.

Note 5.   This table also applies to household cooking appliances rated over 1 ¾ kW and used in instructional programs.

(FPN): See Examples 1(a), 1(b), 2(a), and 2(b) in Tables and Examples.

220-21. Noncoincident Loads.   Where it is unlikely that two dissimilar loads will be in use simultaneously, it shall be permissible to omit the smaller of the two in computing the total load of a feeder.

220-22. Feeder Neutral Load. The feeder neutral load shall be the maximum unbalance of the load determined by this article. The maximum unbalanced load shall be the maximum net computed load between the neutral and any one ungrounded conductor, except that the load thus obtained shall be multiplied by 140 percent for 5-wire, 2-phase systems. For a feeder supplying household electric ranges, wall-mounted ovens, counter-mounted cooking units, and electric dryers the maximum unbalanced load shall be considered as 70 percent of the load on the ungrounded conductors, as determined in accordance with Table 220-19 for ranges and Table 220-18 for dryers. For 3-wire dc or single-phase ac, 4-wire, 3-phase, and 5-wire, 2-phase systems, a further demand factor of 70 percent shall be permitted for that portion of the unbalanced load in excess of 200 amperes. There shall be no reduction of the neutral capacity for that portion of the load which consists of electric-discharge lighting, data processing, or similar equipment, and supplied from a 4-wire, wye-connected 3-phase system.

(FPN): See Examples 1(a), 1(b), and 2(b) in Tables and Examples.

C. Optional Calculations for Computing Feeder and Service Loads

220-30. Optional Calculation — Dwelling Unit.

(a) Feeder and Service Load. For a dwelling unit having the total connected load served by a single 3-wire, 120/240-volt or 208Y/120-volt set of service-entrance or feeder conductors with an ampacity of 100 or greater, it shall be permissible to compute the feeder and service loads in accordance with Table 220-30 instead of the method specified in Part B of this article. Feeder and service-entrance conductors whose demand load is determined by this optional calculation shall be permitted to have the neutral load determined by Section 220-22.

---

best source is directly from the publisher by mail, but electrical suppliers usually have it. The address to obtain the NEC is

National Fire Protection Association
Batterymarch Park
Quincy, MA 02269

The price is $18.50, softbound. A looseleaf edition is $24.50. There is also an abridged edition of the NEC covering One- and Two-Family Dwellings, selling for $13.50. An excellent handbook is also available for $39.50. This Handbook has a number of illustrations and special explanations giving the reasons why certain rules and methods must be used. Following is a price list for your convenience when ordering:

Item No.	Name	Price	Freight
M2-70-87	1987 NEC Softbound	$18.50	$2.85
M2-70-87LL	1987 NEC Looseleaf	24.50	2.85
M2-70A-87	1987 NEC Abridged Ed.	13.50	2.85
M2-SET-87	1987 NEC & Handbook (A good buy)	52.50	2.85

Fig. 11-3. Table 220-30 Optional Calculation for Dwelling Unit (part of Article 220). Article 220-31 Optional Calculation for Additional Loads in Existing Dwelling. Complete. Both excerpted from the 1987 NEC NFPA 70A-1987. Reprinted with permission from NFPA 70-A-1987. Electrical Code for One- and Two-Family Dwellings, Copyright © 1986, National Fire Protection Association, Quincy, MA 02269. This reprinted material is not the complete and official position of the NFPA on the referenced subject, which is represented only by the standard in its entirety.

(b) Loads. The loads identified in Table 220-30 as "other load" and as "remainder of other load" shall include the following:

(1) 1500 volt-amperes for each 2-wire, 20-ampere small appliance branch circuit and each laundry branch circuit specified in Section 220-16.

(2) 3 volt-amperes per square foot (0.093 sq m) for general lighting and general-use receptacles.

(3) The nameplate rating of all fastened in place appliances, ranges, wall-mounted ovens, and counter-top cooking units.

(4) The nameplate ampere or kVA rating of all motors and of all low-power-factor loads.

Table 220-30
Optional Calculation for Dwelling Unit
Load in kVA

Largest of the following four selections.

(1) 100 percent of the nameplate rating(s) of the air conditioning and cooling, including heat pump compressors.

(2) 65 percent of the nameplate rating(s) of the central electric space heating including integral supplemental heating in heat pumps.

(3) 65 percent of the nameplate rating(s) of electric space heating if less than four separately controlled units.

(4) 40 percent of the nameplate rating(s) of electric space heating of four or more separately controlled units.

Plus: 100 percent of the first 10 kVA of all other load. 40 percent of the remainder of all other load.

220-31. Optional Calculation for Additional Loads in Existing Dwelling Unit. For an existing dwelling unit presently being served by an existing 120/240 volt or 208Y/120, 3-wire service, it shall be permissible to compute load calculations as follows:

Load (in kVA)	Percent of Load
First 8 kVA of load at	100%
Remainder of load at	40%

Load calculation shall include lighting at 3 volt-amperes per square foot (0.093 sq m); 1500 volt-amperes for each 20-ampere appliance circuit; range or wall-mounted oven and counter-mounted cooking unit, and other appliances that are permanently connected or fastened in place, at nameplate rating.

If air-conditioning equipment or electric space heating equipment is to be installed the following formula shall be applied to determine if the existing service is of sufficient size.

```
Air-conditioning equipment* .........................   100%
Central electric space heating* ......................   100%
Less than four separately controlled space heating units*  ....   100%
First 8 kVA of all other load  .........................   100%
Remainder of all other load  .........................   40%
```
Other loads shall include:
1500 volt-amperes for each 20-ampere appliance circuit.
Lighting and portable appliances at 3 volt-amperes per square foot (0.093 sq m)
Household range or wall-mounted oven and counter-mounted cooking unit.
All other appliances fastened in place, including four or more separately controlled space heating units, at nameplate rating.

* Use larger connected load of air conditioning and space heating, but not both.

PURPOSE OF THE NEC AND THIS BOOK

With this book, I am attempting to teach you how to complete an electrical installation as follows.

▶ Using correct methods and practices.
▶ Using the proper tools safely and efficiently.
▶ Understanding the *why* of the how-to instructions given.

▶ The calculation of wire sizes and ampacities for the expected load on the system.
▶ Finally and most importantly, installing a completely safe wiring job in accordance with the applicable electrical codes.

The installation of electrical wiring, while not difficult, must be done very carefully and

3. Three-Wire, Single-Phase Dwelling Services. In dwelling units, conductors, as listed below, shall be permitted to be utilized as three-wire, single-phase, service-entrance conductors and the three-wire, single-phase feeder that carries the total current supplied by that service. Grounded service-entrance conductors shall be permitted to be two AWG sizes smaller than the ungrounded conductors provided the requirements of Section 230-42 are met.

Conductor Types and Sizes
RH-RHH-RHW-THW-THWN-THHN-XHHW

Copper	Aluminum and Copper-Clad AL	Service Rating in Amps
AWG	AWG	
4	2	100
3	1	110
2	1/0	125
1	2/0	150
1/0	3/0	175
2/0	4/0	200

Fig. 11-4. Note 3. Three-Wire, Single-Phase Dwelling Services, to Table 310-16, including the table listing Conductor Types and sizes—RH-RHH-THW-THWN-THHN-XHHW. Reprinted with permission from NFPA 70A-1987, Electrical Code for One- and Two-Family Dwellings, Copyright © 1986, National Fire Protection Association, Quincy, MA 02269. This reprinted material is not the complete and official position of the NFPA on the referenced subject, which is represented only by the standard in its entirety.

Fig. 11-5. *This and the following page are excerpts from "Amendments to the NEC" formulated by the City of Troy, Michigan. (Courtesy City of Troy, MI.).*

Secs. 7-35-7.45. Reserved.

ARTICLE III.STANDARDS*

Sec. 7-46. Adoption generally.

Pursuant to the recommendation of the electrical commission of the city the following are hereby adopted as the safe and practical standards for the installation, alteration, and use of electrical equipment in the city. (Code 1953, § 21-501; Ord. No. 64-0-68, § 1, 10-6-64)

Sec. 7-47. Code adopted.

The 1978 edition of the rules and regulations of the National Fire Protection Association for electrical wiring and apparatus contained in the code known as the National Electrical Code approved by the American Insurance Association, except as modified hereinbelow, are hereby adopted and incorporated as fully as if set out at length herein. (Code 1953, § 21-102; Ord. No. 64-0-68, § 1, 10-6-64; Ord. No. 76-0-11, § 1, 2-17-76; Ord. No. 79-0-37, § 1, 8-21-79)

*State law reference—Authority to provide standards, III. Rev. Stats., Ch. 24, § 11-37-3. Supp. No. 34

§ 7-50   ZION CODE   § 7-50

or equivalent and 1 No. 5. Minimum of 1 240 volt circuit and 8 120 volt lighting circuits, all branch circuits to be protected by automatic circuit breaker or fusestate.

(5) Services for commercial buildings using 4,000 watts or more shall not be smaller than three No. 3 wires and 100 ampere service switch. All circuit wiring in commercial buildings shall not be less than No. 12RC wire and shall be either RC or TW Type T. Overload protection or limiting fuses shall be installed in installations over 4,000 watts. House wiring or installations under 4,000 watts will be fused according to NEC standards.

(6) #14 wire in all areas, except kitchen, utility, and dinette which will be #12 wire. Kitchen area, 2 #12 circuits to receptacles. In areas where #14 wire is used, maximum outlets allowed is 10, 1,000 watts maximum per circuit. Separate circuit to heating system with fused disconnecting means within the reach of heating system fused properly.

(7) All underground wiring if not approved direct burial cable must be protected with metal or fiber covering placed at least 30 inches below grade.

Lead covering will not be considered, sufficient protection.

(8) All motors permanently installed shall be wired on a separate circuit with externally operated fused switch as near to the motor as within 5 feet.

(9) All transformers used for neon inside window signs or borders shall be indoor type and shall be enclosed in a grounded metal box. Outside neon signs or borders shall be installed with outside type transformers, weatherproof type.

All leads from such metal boxes shall be brought on through porcelain-glass or other bushings of equal dielectric strength. All high tension wiring for the electric service shall meet the requirements as covered in this chapter; except not to exceed three feet of Greenfield flexible conduit may be used in making connections to transformers. All high tension connections for window signs and borders shall be installed on glass insulators not less than 1½ inches long and securely fastened to the window frame, except where wires are hanging free in air and of necessity across the window. All connections from high tension cables to sign shall be covered with glass insulators of equal dielectric properties and strength.

(10) System or common grounding conductors shall be attached to the street side of water meters, using not less than No. 4 wire firmly stapled to beams, joists or supporting walls from box to connection of ground A driven ground may be used inside or outside the building but directly below the service switch or meter box if no other means of grounding is available. When clamping ground conductor to driven ground or water pipe an approved ground clamp must be used which provides both mechanical and electrical connection.

(11) Electric space heating equipment: Fixed indoor electrical space heating equipment to be listed by the Underwriter's Laboratories, Inc., and have affixed labels of the Underwriter's Laboratories, Inc., as having been tested and approved for such installations. All fixed indoor electrical space heating equipment and installations to comply with Article 422-40 of the latest edition of the National Electrical Code.

(12) Swimming pools: Shall comply with Article 680 of the National Electrical Code of 1978 for all electrical installations and equipment for swimming pools. All underwater lights to be 24 volts or less. (Code 1953, § 21-501; Ord. No. 64-0-68, § 1, 10-6-64; Ord. No. 77-0-9, § 2, 2-15-77; Ord. No. 79-0-37, § § 2, 3, 8-21-79)

Secs. 7-51-7-60. Reserved.
Supp. No. 34

accurately. One wire misplaced or improperly attached can blow up in your face. If you are not prepared to follow rules and instructions to the letter, it would be best to employ a licensed electrician to do your wiring.

In almost every instance, wiring can be done with all power off. If you are hesitant about making a final connection to hot (live) terminals, employ a licensed electrician to make this connection. Before calling the electrician, carefully double check all your work.

UNDERWRITERS LABORATORIES

Underwriters Laboratories, Inc. is a testing facility devoted to testing nearly all manufactured products that have any connection with electricity, burglar protection, casualty and chemical hazard, fire protection, hazardous locations, heating, air conditioning and refrigeration, and marine applications. Any material or product that has anything to do with personal safety has the familiar UL label prominently displayed. This is a pledge of safety of product or material. Look for this symbol when buying any electrical materials and devices. You can write or call at:

Underwriters Laboratories, Inc.
333 Phingston Rd.
Northbrook, Illinois 60062
(312) 272-8880

Chapter 12

Troubleshooting

The process of troubleshooting is much like playing detective. You must ferret out the source of trouble. A blown fuse can be caused by many different conditions. These faults can be localized or affect many other homes in addition to your own.

NO POWER

A condition of no power might affect only your home or the whole area. Look out the window at neighbors homes. If the trouble is general, call the utility. Heavy storms can cause power failures. Utility equipment failure happens occasionally, but it can, in some cases, be automatically bypassed.

If your house has no power at all, there are two causes:

▶ Your main breaker has tripped (or the main fuses have blown).

▶ There are open lines between your meter

and the utility connection, either at the pole or at the service drop where it is attached to your outside wall. Using a flashlight, check the position of your breaker handle. Some breakers have three positions: ON/TRIPPED/OFF, in that order. The tripped position is halfway between ON and OFF. The other type of breaker is either ON or OFF. The tripped position is also the OFF position.

If the breaker has tripped, reset it to ON. If the breaker has the center TRIPPED position, it must be pushed to the OFF position first, then pushed to the ON position. If the handle stays in the ON position, the problem is usually a low voltage condition caused by the utility. The breaker will trip if the voltage is low. The low voltage value causes an increase in amperage *above* the rating of the breaker, and it will activate. This is a random condition and might never happen again. If the breaker will not stay in the

ON position, there is a fault on your premises.

If the breaker stays in the ON position but there is still no power, test for power with a line voltage tester. The breaker panel cover must be removed to make this test. When removing or replacing the cover, be very careful not to permit the corners of the cover to come in contact with live terminals inside the panel. If it is dark in the area, have someone hold a flashlight so you can remove and replace the cover safely.

There is a possibility the breaker is faulty, or wires might have come loose in the panel. Test first where the power enters the panel (usually the top). If you have no power anyplace in the panel, then power is not getting to the service entrance equipment and the utility should be called. Your service-drop wires are connected to the wires on the pole by a split connector. These clamps sometimes break when they are old, and this lets your supply wires hang loose. This is the responsibility of the utility. In daylight, look up at the pole and you might see a wire dangling free. I have had one of these wires come loose and cut off half the lights in a house. Some connectors just work loose and allow the wind to make and break the connection, causing the lights to go on and off. These problems are few but they do happen.

IN-HOUSE FAULTS

Problems encountered in the house are faulty appliance cords and overloaded circuits. Most overloaded circuits are caused by too many appliances being plugged in to an already nearly overloaded circuit. With a fused panel, it is possible to have installed an undersized fuse. This is very unlikely, but check it out.

Circuits using #12 copper wire are to be fused at 20A or be protected by a 20A breaker.

Overloaded circuits can be relieved by moving some heavy appliances to other circuits. It may be that a new circuit needs to be installed. The service equipment must be large enough to handle this new addition. Kitchen appliances now are drawing more current as new models are designed having higher wattage than last year's models. New percolators now draw 1000 watts.

If you plug in an appliance and the plug sparks or the cord flames up, the problem is obvious. This type of fault will show as a blackened window in the fuse. An overload will not show this; the fuse link in the window will have disappeared (it melted and fell down inside the fuse body). With a short circuit, the fuse really "blows."

With branch circuit fuses and breakers, if the fuse blows again or the breaker will not stay on, unplug all appliances. Then reset the breaker or replace the fuse. If either holds, one of the unplugged appliances might be at fault. Carefully plug in each appliance in turn. Plug in only one appliance at a time. At each plug-in, if the breaker holds or the fuse holds, unplug that appliance. Do this for each appliance in turn. If no appliance opens the circuit, then the appliances are ok.

It could be that the total load of all appliances together on the circuit is causing an overload. Two appliances at 1000W can draw 17.4A; that is more than a 15A circuit will handle. A modern kitchen circuit installed correctly can carry a load of 20A. This should not be for a long time as circuits are derated to further protect the wire and insulation. A load is considered constant if it is connected for longer than 3 hours.

DEVICE FAILURES

Devices such as switches and receptacles

wear out from daily use. Switches get high usage. Receptacles get hard usage, especially in the kitchen or where vacuum cleaners are plugged in. Long cords are stretched and pulled on the receptacle contacts, as do those cords on portable electric tools. This strain does wear out and break receptacles. Because these devices are reasonable in cost, buy the best you can afford. Better-quality devices last longer and give better service.

Always turn off the power when replacing any electrical device. Remove the fuse that controls the circuit feeding the device to be replaced. Put the removed fuse in your pocket to prevent others from replacing it. The cardinal rule when working with electricity is to treat every wire as hot (live) unless you can see both ends and the full length between them. Just be careful! Turning a breaker off poses some risk after you start to remove the device. As a precaution, post a large sign covering the whole breaker panel to alert others: "DO NOT TOUCH! WORKING ON THE WIRING."

REPAIRING THE DEFECTIVE APPLIANCE

Appliances, including lamps and power tools, blow fuses and trip breakers usually because of cord defects. Constant use flexes and wears the cord and the attachment plug on the end. Cords can be repaired under some conditions. *Do not splice a cord.* The splice is usually unsatisfactory and the tape never stays tight. It is less expensive, faster, and safer to replace the whole cord. Worn cord ends can be cut off and the cord can be reattached to the appliance. Do the same with the plug end by putting on a new attachment plug.

Buy good-quality cords and accessories. Cords are available in various lengths with the attachment plug molded on. Lightweight cords are good for table and floor lamps. Toasters and similar high-wattage appliances require heavy-duty cord with No. 16 wire such as the SPT-2 type. These heavy-duty cords are available with molded-on plugs. This heavy cord can be used for all heavy-wattage kitchen appliances. SV cord is a special type made only for vacuum cleaner use.

Newer, light-duty attachment plugs are made with no exposed screw terminals to contact metal receptacle cover plates and arc. These plugs, now required by the NEC, open to expose inside terminals for connecting the wires from the cord. There are heavier duty plugs for use on power tools and extension cords. Because these cords are larger diameter, larger-size plugs are required to accept these heavier cords.

BALANCING THE LOAD BETWEEN CIRCUITS

Modern houses have 240V service. This consists of two hot wires—one red and one black (sometimes two black)—and one neutral. The neutral wire can be either bare or insulated. Service-entrance cable (SE) has a bare neutral wrapped around the two insulated wires. Wires in conduit have three insulated wires. The two hot wires have a voltage to ground of 120V. They also have a voltage to the neutral wire of 120V. Between the two hot wires there is a voltage of 240V. Voltage can vary with the time of day and for other reasons. Plus or minus 5 volts is the normal variation. This 240 volts supplies such fixed appliances as electric ranges (or cooktops and wall-mounted ovens), electric dryers, and electric water heaters. Motors running on 240V also are fed by the two hot wires only; they do not need the neutral wire.

Because lights, small portable appliances, vacuums, and many power tools run on 120V minimal voltage, the neutral is needed in the distribution panel to provide the necessary 120V. As an example, if a toaster is connected to one side of the 240V wiring through a receptacle, one wire goes to the neutral wire in the panel (through the house wiring), and the other wire goes to one hot wire. In this case, there will be a current flow in the panel neutral. This is what the neutral is for. However, a *large* current flow in the neutral is undesirable and should be minimized. Hence, you arrange to equalize the current draw on both sides of the panel; then there will be a minimum of current flowing in the neutral. The ideal condition would be for no current flow in the neutral, but this can never be attained—thus the neutral.

Add up all the loads expected in each circuit and try to balance current flow on both sides. Some circuits might have to be moved to the other side, but exact balance is not necessary.

ADDING TO PRESENT WIRING

Houses built in the last 20 years usually have enough electrical capacity for your present needs. Small, older houses might not be wired for 240V because in former years, heavy-current-usage appliances were only provided with 60A service-entrance equipment because that capacity was enough. The major load might have been an electric range (instead of perhaps a gas range), a radio, toaster, and an iron.

At one time, many utilities supplied electricity to the electric water heater directly for a set monthly fee based on the heater size. In such a case, the electric heater was supplied independently of the main service-entrance equipment. Therefore, the service-entrance ca-

ble sizing calculation did not take into account this additional load.

The National Electrical Code now requires 100A service-entrance conductors and equipment for any net computed load over 10 kW. Existing dwellings can continue to use 60A service.

As an example of load calculations, assume the following. A dwelling has a floor area of 900 square feet exclusive of basement and attic (unusable). See Table 12-1.

Because the net computed load exceeds 10 kW (10,000 watts), you must have an ampacity of 100A. No. 3 AWG cable can be used. The neutral size is computed at 70 percent of the ungrounded wires' ampacity. Service-entrance cable will have the correct ratio of grounded wire size to the two ungrounded wire sizes. If you are using conduit from the service head down to the meter, buy enough single wire for the two wires needed. For the neutral wire, buy one length. For 100A service, the hot wires should be No. 3 and the neutral should be No. 4. There is no No. 5. The inspector might approve No. 6 for the neutral; ask.

All three wires can be the same size. Be sure to paint both ends of the neutral white (white tape can be used also). The utility requires that three or four feet of each of the three wires extend from the service head for connection to their service drop at the building wall. Service-entrance cable can be used instead of conduit above the meter. I recommend a service head on the cable because it looks much better.

In Fig. 12-1, original wiring is shown from a 1942 house. The wiring is the minimum required by the FHA at that time. Each room had two receptacles and a switch-controlled ceiling fixture.

There was no automatic heating system,

```
General Lighting Load
    900 sq. ft. @ 3W per foot.................2700W
Minimum number of branch circuits required
General Lighting Load
    2700 ÷ 120 = 22.5A: or two—15A 2-wire circuits
                        or two—20A 2-wire circuits
Small Appliance Load: two 20A 2-wire circuits
                      one 20A 2-wire circuit

Minimum Size Feeders Required
  Computed Load
    General  Lighting........................2700W
    Small Appliance Load.....................3000W
    Laundry Load.............................1500W
      Total (less Electric Range)............7200W
      3000W @ 100%...........................3000W
      7200W − 3000W = 4200 @ 35%.......1470W
    Net Computed Load (less range)........4470W
    Range Load.............................8000W
      Net Computed Load, with Range....12,470W

For 120/240V 3-wire system feeders: 12,470 ÷ 240 = 51.96
```

Table 12-1. Computed Load

only a coal-fired gravity furnace. We had a refrigerator, iron, toaster, radio, and later, a power saw in the basement. The basement plan shows four 15A circuits originating at the 60A service-entrance panel. The basement lights were four pull-chain porcelain sockets. There was a light at the foot of the basement stairs, controlled by a switch at the first-floor level. The laundry had an outlet on the pull-chain porcelain socket. These are still available, but the outlet is now grounded. No convenience receptacles were provided in the basement. See Figs. 12-1 and 12-2.

Refer to the modernized wiring plans shown in Figs. 12-3 and 12-4. Notice that the service-entrance panel is increased to 125A. (Perhaps 100A would be sufficient also.) Receptacles in the rooms are installed to conform to the present code. The wiring is all new because the

original Romex was ungrounded and provided no grounding for the metal frames of appliances, etc.

In the kitchen there are two small appliance circuits, and the range circuits are designated as D and R. Four new circuits were added: workshop, range, heating/ac, and laundry. This shows on the new basement plan (Fig. 12-3). The original three other circuits were kept but were rewired with new cable. All of the wiring is new.

This type of project is a major undertaking and takes a great deal of time. Make a wiring layout to take to the city inspector, and ask for help in guiding and advising you. With the help of the current *1987 National Electrical Code (NEC)*, you can do a fine job. If you now live in the house, do the job piecemeal. If the house is vacant, start with the service-entrance equip-

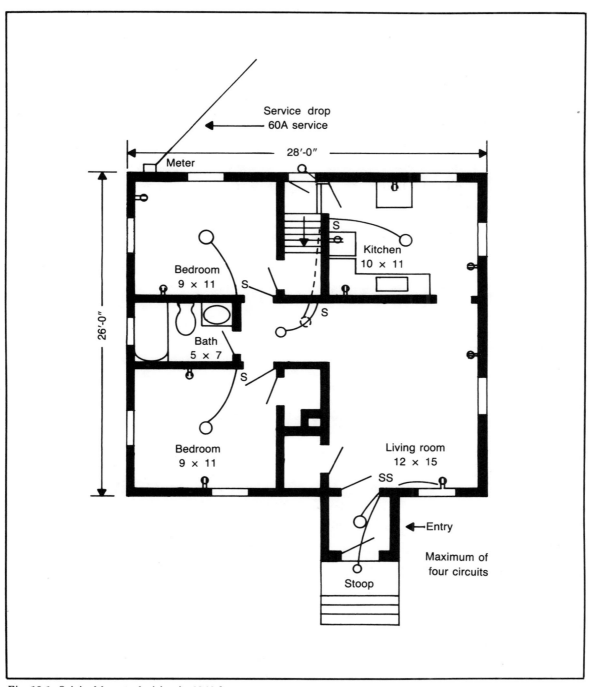

Fig. 12-1. Original layout of wiring in 1940 house.

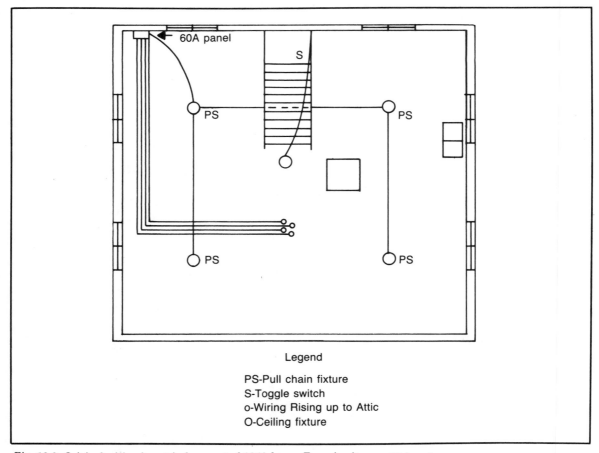

60A panel

S

PS

PS

PS

PS

Legend

PS-Pull chain fixture
S-Toggle switch
o-Wiring Rising up to Attic
O-Ceiling fixture

Fig. 12-2. Original wiring layout in basement of 1940 house. Four circuits were 15A each.

ment and do the whole job at your convenience. Borrow power or install a temporary power setup.

SMALL APPLIANCE REPAIR

Start with easy repairs and you will learn how to do them properly and accurately. A simple repair is the replacement of the socket and cord on a table or floor lamp. Generally, if the socket is defective, the cord is also worn and should be replaced. Two common places of wear are where the cord enters the lamp base and

at the connection to the attachment plug. It is best to replace the whole electrical assembly at one time. Cord is sold by the foot or as an assembly of cord and plug. If the plug looks sturdy, buy the assembly available in different lengths. Do not buy too long a length. Excess cord length gets stepped on and damaged.

To repair a lamp, see Figs. 12-5 through 12-7. Unplug the cord (this eliminates the sparks) and remove the shade. The top part of the *harp* (the metal loop that supports the shade on some lamps) can be removed because its

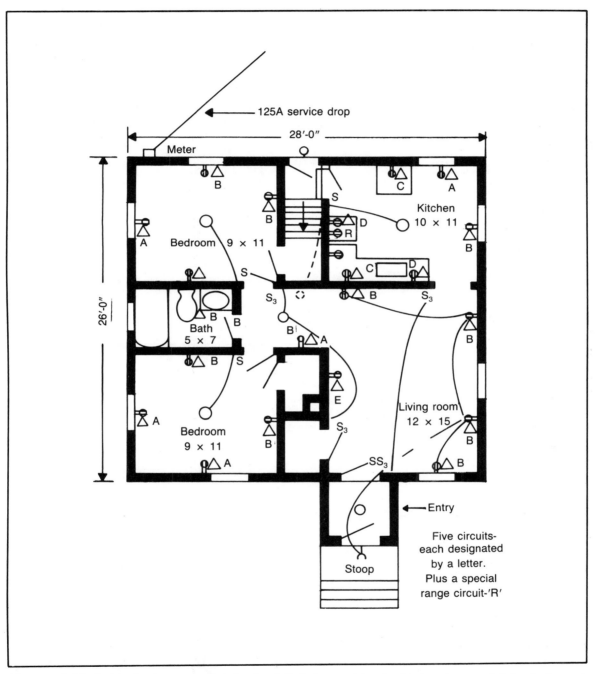

Fig. 12-3. Original wiring has been removed or abandoned. Entirely new wiring plan has been drawn up to bring the house up to the 1987 NEC.

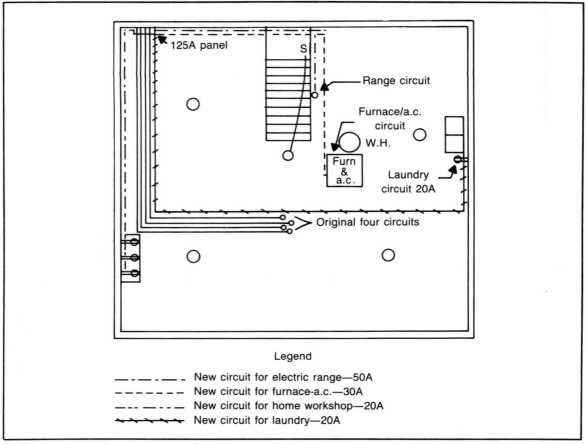

Fig. 12-4. Basement plan showing increased service-entrance equipment and addition of four more circuits. Water heater and furnace are gas-fired.

lower ends fit into two sockets. On others, the complete harp is in one piece.

If the harp is not removable, proceed to dismantle the socket. Unscrew the bulb. Next to the switch on the socket (whether it is push-button or turn knob) is the word "push." Push on this word while trying to tilt the socket shell away from you slightly. At this point, the socket shell should part from the socket base. If the shell does not separate from the base readily, insert a small pocket screwdriver between the

shell and the base. By gently prying outward on the screwdriver handle, you are actually "pushing" on the shell. This will release it from the base. (It's simple to do, but complicated to describe.)

Now push some of the cord back into the lamp base for enough slack to lift out the socket far enough to disconnect the wires from it. If you go this far, it is best to replace everything—cord, plug and socket. About $4 should do it. The socket base (bottom half of

(for three-way light bulb). Any of these three types of sockets work on most lamps. As a precaution, take the old socket shell to compare the threads (just to make sure). Wasted trips are wasted money.

To replace the attachment plug on a lamp cord, extension cord, or power tool is simple to do, but it must be done right. The plug might be broken or the cord is frayed at the plug end. Old plugs that have seen much use should be

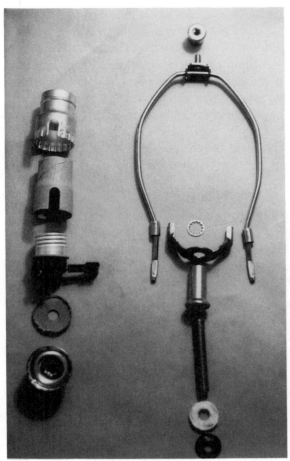

Fig. 12-5. Table lamp socket and harp disassembled. Parts on the left are, from the top: Shell, fiber insulator, three-way socket, and lower insulator for shell bottom. On the right are: harp, lock-washer, spacer, washer, 1/8-inch running thread nipple (used in lamp and fixture assembly), felt washer, and jamb nut.

Fig. 12-6. Socket ready for assembly. Notice the "Underwriter's knot" below the socket to prevent strain on the wires and terminals.

the shell) is threaded onto a 1/8-inch pipe nipple. Pipe and pipe nipples are designated by the internal diameter. Matching fittings are designated by the nipple size, called *nominal*.

There might be a small setscrew in the side of the threaded part that keeps this base from turning. The base also secures the harp. Buy the complete socket. Match the operating type—pushbutton, on-off switch, or three-way

Fig. 12-7. Method of removing socket shell from base cap.

replaced. The new type of attachment plugs are designed so that no bare wire ends or screws are exposed to contact a metal receptacle cover plate. This is called a *dead front* type of plug. The 1987 NEC, Section 410-56(d) requires this type as a safeguard to prevent fires. This new design costs about $1.50.

I have seen many metal receptacle cover plates with burns on them from exposed terminal plugs that were supposed to have a fiber washer fitted over these exposed terminals, but usually the washer is missing just when it's needed. It is common for receptacles to have

metal plates; they are even used in homes instead of plastic ones. Perhaps this is the reason for the new design attachment plugs.

Shop clearance counters at hardware and home improvement centers often have astounding bargains in electrical supplies. Even if the items are reduced in price, check other stores before buying. Also check the items for perfect condition and the privilege of return if defective in any way. For cord repair, see Figs. 3-18 through 3-23.

Don't buy just because items are reduced. Bargains are to be found in cable reel ends or for other reasons. Just make sure the lengths are long enough for your needs. Too short a length will do you no good. When measuring for length, allow for the 3 feet at the service-entrance head and at least 2 feet at the panel end for service-entrance cable and other expensive cable that you will need.

Different types of attachment plugs require different connection methods. Some plugs for use with "rip" cord (the common lamp and household extension cord type) just need to have the square cut end inserted into the side of the plug body and the lever pushed down to make contact. These plugs are not very satisfactory because of the poor contact made between the cord and plug prongs (points puncture the cord insulation and "sort of make contact"). Buy the new dead front type; it's worth it. Illustrations in this chapter show connections being made to various plugs and cord connectors and a trouble light is wired to a cord. All of these wiring methods are shown using three-wire cord.

Cords on power tools are replaced in a similar manner. Also illustrated is the repairing of the connection to a trouble light head (McGill Manufacturing Co. is a top-quality brand). Although not necessary, crimp-on terminals have been put on the wire ends of the cord itself. This is not necessary, but desirable. Stranded wire tends to spread when tightened under terminal screws; crimp-on terminals eliminate this. The heavy-duty cord is similar to cord used on power tools. Black, white, and green wires are attached to the brass, silver and green terminal screws. There are plastic ribs between the terminal screws (barriers) that prevent stray strands of wire from shorting over to the adjacent terminal. The crimp-on terminals are bare. There are also those with a plastic sleeve over the crimp-on part. These are crimped directly over the plastic using a special crimping tool.

The cords on the new double-insulated power tools should be replaced *only* by the service center for that tool. This is because special equipment is used to maintain the integrity of the insulation system.

LOCATING THE CAUSE OF POWER FAILURES

If a fuse blows or a breaker trips and you are using a power tool or appliance, immediately disconnect the item and any lamps, and turn off any ceiling lights. Inspect the panel. A fused panel suffering from a simple overload will show a ''missing'' fuse link that normally shows in the fuse window. A short circuit shows a darkened window caused by an arc. *Arc*, as in arc welding, is an actual flame that carbonizes the fuse link, making the dark-colored window. In this manner, you can look for trouble more easily. Circuit breakers do not tell you anything except that the circuit has been protected as required. The circuit breaker trips whether it's an overload or a short circuit.

If many appliances were being used on the circuit, then removing one or more and plugging them in on another circuit might solve the problem. Check all cords on anything used on this circuit. Any cord fault might cause the trouble. I have a toaster that pulls (uses) 10.5A and a vegetable juicer that pulls 6.7A. This totals 17.2A. Because these both are on a 20A circuit, this is allowable. No single portable appliance over 16A (1840W) can be plugged into this circuit. So my two heavy amperage appliances are OK to use on this circuit.

In a modern house, it is unlikely that the building wiring is defective, but it is still a distinct possibility to be considered. At this point, with everything unplugged, reset the breaker once. If the breaker will not stay on, take your light bulb adapter (socket-to-prongs, explained in Chapter 3) and try all the receptacles for power. Something might still be plugged in. If now the breaker stays on, plug in one appliance at a time.

If the breaker holds (stays on) remove the appliance. Now do the same with each appliance in turn, noting whether or not it was the cause before you plug in another one. If you have help, let the person stand near the panel to call out if the breaker trips. If the breaker holds for each appliance in turn, add up the combined amperage (watts × 120V = amperes). You might have a borderline overload circuit.

If you have a low voltage condition from the utility, the appliances and other items will draw more amperes and thus overload the affected circuit. This entire procedure explained above is exactly the same for a panel using fuses. The difference is that spare fuses must be on hand to replace the blown fuse instead of resetting the breaker. Spare fuses should always be on hand in any case.

Another possible cause of trouble might be defective or worn devices. Switches or receptacles that have heavy usage can cause a fault current. This will cause the breaker to trip or the fuse to blow. Feel cover plates at switches, receptacles and junction boxes. Sometimes this can alert you to trouble spots. Now remove the plates and inspect the inside with a flashlight and your nose. Not too close with the nose! It is easy to detect burned odors by the smell. Just don't get too close that you get a shock; keep 4 to 6 inches back.

Other areas of trouble might be in junction boxes in the basement or attic. Loose wire nuts or defective soldered wire connections can heat up. The soldering of electric wires is an art that has to be practiced to become proficient at. Be sure to check wire nuts and soldered connections with the power *off!* As described in Chapter 1, check and tighten all terminal screws found in the service-entrance/distribution panel. Any sustained current passing through wires and terminals causes some heating. This sustained heating eventually loosens connections. This is why all terminal screws must be kept tight. Hazardous conditions such as these cause many house fires every year.

Even though you have eliminated the condition causing the breaker to trip or the fuse to blow, it is wise to tighten all terminal screws on switches and receptacles. This is especially true if the house is old. It takes only a few minutes to check each device. First turn off the power to the circuit you will work on. Check for voltage *first*. Remove the cover plate and the two screws holding the device. Pull the device out far enough to reach the screws and tighten them. Replace the device and go on to the next one. Do all devices on one circuit, but be sure to check for voltage before you touch any terminals.

Most circuit faults are not in the wiring itself. Look at the appliances first before disconnecting any wiring. Many split-phase motors on furnace blowers and oil burners have an internal switch inside the motor housing. This switch becomes covered with an oil film from oil in the bearings. Because oil attracts dust, these contacts tend to become dirty and in some cases do not make contact with each other. In such a case, the motor will not start, but it will keep trying until it trips its own overload protection. If this overload does not shut off the motor circuit, the fuse or breaker in the distribution panel will disconnect the circuit. Furnaces are to be on a separate circuit, but other equipment might have been connected to this same circuit. This will cause a localized power outage. Many of these motors have an automatic overload that reconnects the motor. This has the effect of letting the motor try to start many times (eventually blowing the fuse or tripping the breaker). See Figs. 12-8 through 12-10 for motor repair.

These motors have two separate windings (wire coils) inside, the start winding and the run winding. Both windings are ready to start when the motor is at rest. When the motor receives power, both windings combine forces to turn whatever load is connected—pump, fan or whatever. (If the centrifugal switch sends power to the start winding to help the run winding get the motor turning, that is fine.)

If the switch contacts are dirty, the start winding cannot help as it is not powered. In this case, the motor will keep trying to start and finally blow the fuse or trip the circuit breaker. Under normal conditions, the start winding provides the extra torque to start the motor. When the motor gets up to its speed (usually 1825 RPM), the centrifugal levers on the motor shaft operate to disconnect the start winding by opening the centrifugal switch. The motor then runs on the run winding. When the motor stops running (is shut off manually or automatically), the centrifugal start winding switch closes, ready for the next start.

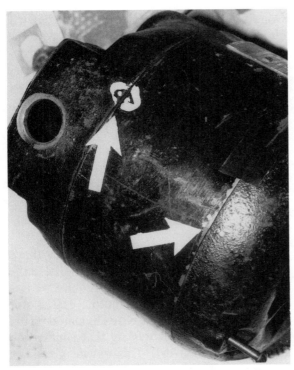

Fig. 12-8. Motor with centrifugal switch inside end bell. Make punch marks on end bells and directly opposite on the center housing. Mark one end with one mark on each end bell and housing; mark the other end with two marks on each end bell and housing. This ensures correct re-assembly. Note the through-bolt, lower right, partially removed.

tor end bell off. This is the end having the connections to the power supply. These terminals have a cover plate over them to be removed if you have to disconnect the motor to work on it. The centrifugal switch is mounted on this end bell. Figure 12-9 shows this switch with an arrow pointing to it. The part on the shaft operates the switch by centrifugal force. These switches can be replaced if necessary. When buying, take the old part with you for exact replacement. Repair centers and motor repair shops have parts. Take the motor with you or copy the nameplate information to have with you.

Motors are assembled with four *through bolts* that go through the complete motor housing, holding the two end bells to the center sec-

Other loads might be on this same circuit and will be without power. This is a violation of the NEC because furnace and air conditioning circuits can have no other loads connected to them. In your preliminary inspection, you should have noted this and planned to remove other loads from this circuit. This drawn-out description of motor problems applies to any residential motor except air conditioning compressor motors (a special case). Motors can be oil burner, fan, water pump, sump pump, and the air conditioner fan motor.

Faulty centrifugal switches can be cleaned, if the points are not burned, by taking the mo-

Fig. 12-9. End bell with centrifugal switch, Arrow points to switch contacts. Centrifugal mechanism is still on shaft.

139

Fig. 12-10. Arrow points to self-resetting overload. Sometimes these fail and need to be replaced. Motor parts are available at appliance parts stores.

tion (refer to Fig. 12-8). Before dismantling the motor, mark the end bells so they can be replaced exactly as they were previously. Do this by using a prick punch to mark each end bell and its housing end with adjacent punch marks on housing and end bell.

Each end bell should have different marks: one end to have only one mark each on end bell and center housing, the other end to have two marks each on its end bell and center housing. These end bells fit snugly to assure alignment of the bearings.

After removing the through bolts, tap the end bell free using a screwdriver and hammer.

Do this gently! At some place, the end bell will be slightly raised along its edge. Tap with the screwdriver at this point. Tap all around the circumference to remove the end bell; do not pry on one side only, as you might break it. Remove only the end bell with the centrifugal switch. When reassembling, align the punch marks and tap the end bell all around the edge. Be careful to tap gently because the end bell is a casting and can break. Insert the through bolts before completely seating the end bell. After assembly, try to turn the shaft to make sure it turns free. Double check everything thoroughly before you reconnect the motor.

Special heavy-duty motors (such as those used on water pumps and other uses where the motor must start under a load) require what is known as a *capacitor* type motor. This motor has one or two cylindrical cases mounted on the outside of the motor housing that contain the capacitor (condenser). This device gives the motor a much stronger torque to overcome the heavy load imposed by the pump or large fan that it has to start.

These capacitors sometimes short out and fail, blowing a fuse or tripping a breaker. Unscrew the capacitor case and check and smell the capacitor. You might have to unsolder it and take it to a motor repair shop to have it checked. The rating of the capacitor is printed on its metal or plastic case. After doing any repairs, recheck everything you have done before using the equipment. Be safe!

BUILDING A BATTERY-POWERED TESTER

A simple battery-powered tester can be made by using a 6-volt lantern battery and a door buzzer. Two wires are attached to the battery terminals (you might have to solder them). A length of lamp cord is good. One wire of the cord goes to one battery terminal and the other wire goes to one terminal of the buzzer. A short, single wire goes from the second battery terminal to the second buzzer terminal. The cord should be about 5 or 6 feet long.

At the free end of this cord, separate the two wires for about 2 feet to form test leads. Strip each end 1 inch to form test prods. You might want to solder these ends to make them rigid. As the voltage is only 6 volts, it is not important (but it might be handier) to make the ends rigid by soldering.

Touching the two ends together should make the buzzer sound. This enables you to check any dead circuit (deenergized). This tester cannot be used on live circuits under *any* circumstances.

Using The Battery-Operated (Continuity) Tester

Suppose you suspect a length of Romex has a defect or break in it, or the three wires might have been shorted out due to the cable being crushed at some point. At each end of the length of cable, strip the wires so that you can test them individually. Using your homemade tester, touch one test probe to the black wire; then touch the other probe to the far end of the same black wire. A *buzz* means that the black wire is ok.

Now remove the second probe from the far end of the black and touch either end of the white wire. A buzz indicates a "dead short" between the black and white wires.

Now check the white wire as you first tested the black wire. The white wire could be broken or touching the black wire at a defect area of the cable. Also check the black wire to the ground wire. A buzz here shows a serious

defect in the cable length. You will have to visually check the outside of the cable inch by inch to detect any obvious damage. Some object has crushed the cable even though it might not show. If the defect cannot be found, the cable must be discarded anyway.

If you cut the cable in two, one-half might test ok. You can use this half if you are sure you have tested properly. If you are careful when installing cable and do not pound the staples too tightly, you should have no trouble with new cable.

USING THE LINE VOLTAGE TESTER

To test live wires, a line voltage tester is needed. This can be a $1.49 pocket tester or the professional $25 one. Either will work fine. Just be certain that your fingers do not contact any hot wires. Work carefully and slowly when testing hot wires!

A homemade tester for 240V *and* 120V is easily made from two rubber pigtail sockets. The 6-inch leads have the ends bared about ½ inch. One lead from each socket is cut short, and these two leads are connected together using a wire nut. This connects the two sockets in series with each other, meaning that the electricity must travel through the first lamp and then through the second lamp; thus the tester can be used on 240V or 120V, even though the lamps used are rated for 120V. Build this tester carefully, use it properly, and it will serve you well. Line voltage testers are used to determine if voltage is present. Your homemade tester shows you if the voltage is 120V or 240V by the brightness of the lamps. Half brightness shows 120V, full brightness shows 240V.

Line voltage testers can test for open circuits. They can be used to test both fuses and

141

circuit breakers in place. These testers also test for voltage at receptacles and fixtures. Any other situation where it is necessary to determine the presence of power requires a voltage tester. Low voltage testers are also made for 6V to 50V. These are used to test doorbell circuits, automotive circuits, and furnace and air conditioning equipment *low* voltage circuits.

Continuity testers are for testing fuses and electrical controls *not* in the circuit or with the circuit *disconnected* from the power supply, such as in your hand or on the work bench. These testers only determine whether the circuit is complete or that a control is making or not making contact. *Caution:* Never use this type of tester on live circuits. The tester will be ruined!

Chapter 13

Wiring Materials and Standard Wiring Methods

When doing electrical wiring, it is necessary to use wire, wiring devices, conduit, cabinets, and fastening devices to complete the installation. In addition, construction methods, developed over the years to facilitate the work and make for a sound, safe installation, are explained in this chapter.

WIRES AND CABLES

Electric wire comes in various sizes and forms. The most common types used in house wiring are BX and Romex. Romex is a plastic-covered cable having two or three insulated wires and a bare grounding wire. BX serves the same purpose and has the same construction, but the covering—instead of being plastic covered—is a metal, spirally wrapped covering. This metal covering is more resistant to damage than Romex. For this reason, many areas require BX in all installations.

Other electric wire forms are single wire that is either solid or stranded. Solid wire is one wire covered with insulation. Stranded wire is a number of fine wires twisted together to form one conductor. This type is found in the lamp cords and extension cords. Wire sizes vary from very small to very large. The sizes used in house wiring range from No. 14, for branch circuits, to No. 2/0, used for the wires coming into the building.

Romex

Romex is the most widely used wiring. See Fig. 13-1. The following information also applies to BX, with the exception of the metal covering. (Figures 9-24 through 9-26 illustrate how to cut BX.) Cable sizes commonly used are Nos. 14, 12, and 10. Nos. 14 and 12 are used in branch circuits such as lighting and appliance circuits. No. 10 is used for circuits supplying clothes dryers and electric water heaters. Nos. 8 and 6 are used to supply electric ranges. All

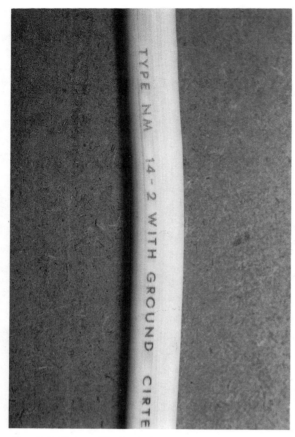

Fig. 13-1. Non-metallic sheathed cable (Romex). Notice the marking on the cable sheath. "Type NM 14-2 with ground Cirtex plastic 600 volts Cerro".

these wire sizes are made in copper and aluminum. Do *not* buy or use wire made of aluminum! Aluminum wire used in homes built during the 1960s caused much damage to the devices used with it and also caused house fires.

Both Romex and BX cables must have a grounding wire included in the cable. This wire is a safeguard for persons and equipment (especially for persons). There might still be Romex on the market without this ground. Do not buy this type of wiring. The correct marking on Romex cable is "(UL) Type NM-12-2 AWG

with ground 600V." This means that the cable has two No. 12 insulated wires and one bare grounding wire. The wire has sufficient insulation to be used on 600 volts. The type NM cable can be used in dry locations only. Type UF may be used in damp but not wet locations (UF may be used also in dry locations). In this respect, it should be stated that if you are doing much wiring (or rewiring) consider buying one size of Romex, namely No. 12-2. It will be cheaper to buy one large coil of one size than two smaller coils. The standard coil of Romex (or BX) is 250 feet, and in this quantity is cheaper by the foot.

A new type of non-metallic cable is designated as NM-B 14-2 600V with ground and NM-B 12-2 600V with ground. This cable has the new 1987 NEC rating of 90° C, allowing it to be used in higher ambient temperatures.

Anchoring Cable to Boxes

Connectors are used to anchor cable to boxes. Previously, separate clamp type connectors were used, having locknuts to hold them in knockouts in the boxes. Boxes are now available with clamps inside the back of the box. These line up with knockouts and when the knockout is removed and the cable is brought through the knockout opening it goes under the clamp and is anchored tightly. The top and bottom of the box each have two knockouts available for use with their built in clamps.

Plastic boxes are now available with or without cable clamps. It is still necessary to staple the cable within 8 inches of the box. This is where the wiring is classed as "new work," as when the studs do not have the wall material applied to them. In remodeling work (old work), the cable cannot be so anchored.

All electrical connections must be made inside outlet or junction boxes. Cable must also be run so that it is protected from damage. (Figures 13-17 and 13-29 illustrate these type of installations.) Cable must be run through bored holes in joists or studs. Exposed joists such as in basements may have cable run on the bottom edge if there are wood strips on each side or under the cable for protection. In walls to be finished with drywall or another covering, cable can be run in notches cut in the face of the stud, but the cable must be protected by a metal plate, 1/16 of an inch thick, fastened over the cable lying in the stud notch. (See Fig. 13-29.)

In attics within 7 feet of the access opening, the cable runs must also be protected by running boards. For anchoring cable, there are available one or two hole straps for use with nails, two-point staples with a flat top to prevent cable damage, and special staples with a plastic top-part, having one nail through each end, to protect the cable from damage when nailing.

Cable must be anchored every 4½ feet. Cable fished inside walls is exempted from this requirement. If a new outlet box or ceiling box is installed, any openings between it and the plaster wall must be patched so that any possible sparks have no chance of getting inside the wall. This is a code requirement in all cases.

Metal boxes used with cable must be grounded to the equipment ground. Cable designated as ''with ground'' must always be used whether you are using metal or plastic boxes. The bare grounding wire in the cable must be attached to the metal box and to the green grounding screw on the device mounted in the box. Nearly all devices now available have this green hex head grounding screw on the device mounting strap, which effectively grounds the metal non-current-carrying strap. Cheaper toggle switches do not.

Armored Bushed Cable (BX)

This type of cable is commonly known as BX. BX gives better protection than Romex, but it is more expensive. Where the use of BX is mandatory, Romex may not be used. The grounding method uses a very narrow aluminum strip that is bent back over the outer metal armor and clamped tightly by the connector either inside or outside the metal box. BX must always be used with metal boxes.

Use a hacksaw to cut BX. Do not cut the armor at right angles to the length of the cable, but rather at right angles to the armor strip. Mark the length needed, plus allow 8 inches inside the junction box to make connections.

Take the marked place in your left hand; step on the cable with your foot allowing enough length to bring the cutting mark waist high. Flex the cable enough to make a hump at the mark while keeping the cable taut. Cut almost through the armor. Flex the cut point till it breaks loose and then cut the wires with diagonal cutters. To bare the wires, back up the 8 inches excess and repeat the sawing. See Figs. 9-24 through 9-26.

Be extremely careful this time not to cut into the wires. Remove the armor by bending and twisting. Prepare the other end of the cable length if you have not already fastened it in place. This method of cutting can also be used in a run of BX already installed in joists or studs. As the cable is already anchored, no foot holding is needed.

Use the same hand method to arch the BX for sawing. Because BX has the steel outer covering that has a sharp cut edge, it is necessary to insert a plastic or fiber bushing between the outer steel casing and the wires inside. This is required by the NEC and will be checked by the inspector. All BX clamps have slots for the purpose of determining the presence of the fiber bushing without dismantling the connection.

SERVICE-ENTRANCE CABLE (SE)

Service-entrance cable is similar to Romex except that the wire is type XHHW (moisture and heat-resistant insulation) and there are two insulated wires, red and black. The neutral wire consists of many fine strands of tinned copper wire wrapped spirally around both insulated wires. This is then covered with a tough outer insulation that is damage resistant. There is no grounding wire as such. The neutral is also the grounding wire. See Figs. 1-2 and 1-3.

The cable brings power through the meter and into the building to the service-entrance equipment. The fine, tinned strands of the neutral are bunched and twisted together and connected to the neutral bar, to the grounding wire in the panel, and in the meter base.

Modern homes usually need large service-entrance cable size such as No. 4, 3, 2, or 1 for 100A, 115A, or 130A capacity respectively. Service-entrance cable may also be used to wire ranges and clothes dryers using No's. 6 and 8 respectively. Watertight cable connectors must be used in wet locations such as at the top entrance to the meter base outdoors to prevent moisture entry. A special service head is used at the point where the service drop connects to the cable. You can purchase service-entrance cable by the foot at home improvement centers. Measure carefully so as to have enough length. This cable is very expensive. Cable with aluminum wires is also satisfactory for ranges, dryers, and electric furnace uses. See Chapter 7 for explanations on the various types of conduits.

SURFACE RACEWAYS

As defined in the NEC, a raceway is designed to hold wires. A *surface raceway* is used for an exposed extension to an existing concealed wiring installation because of building construction or other reasons. Many times it is impossible to conceal needed additional wiring. See Figs. 13-2 through 13-7.

The most popular and perhaps the oldest surface raceway system is that manufactured by the Wiremold Company. Their system is complete; it furnishes not only the raceway, but adapters to change from other raceways (Thinwall, conduit or Romex) to Wiremold. There are special systems that have been developed by the Wiremold Company. These include receptacle strips, Plugmold, and other specialized systems. New kits for do-it-yourselfers are now available also.

Wiremold raceway sizes start at ½-inch wide by $^{11}/_{32}$-inch high to ¾ inches wide by $^{21}/_{32}$ inches high. Special shapes are used for over the floor uses such as in offices where the permanent receptacles are not in convenient spots. This style has sloping sides to minimize tripping. While Wiremold raceways are designed to be bent, with the variety of fittings for use with them, you might not have need to bend the raceway. There are a variety of tools for use with Wiremold, but you most likely will not need them for the amount of work you will be doing.

There are many uses for Wiremold in the home, even though it was originally designed for commercial use. The raceway is very neat in the smaller sizes. It is enameled in a tan color and is quite attractive when installed neatly. You might like to use it in the basement instead of Thinwall, as it gives a finished appearance. Switches, receptacles, ceiling boxes, and many other special adapters make a very neat installation. The Plugmold 2000 raceway is useful for the workshop over the workbench. For the kitchen, there is a stainless steel style with outlets every 6, 12, or 18 inches.

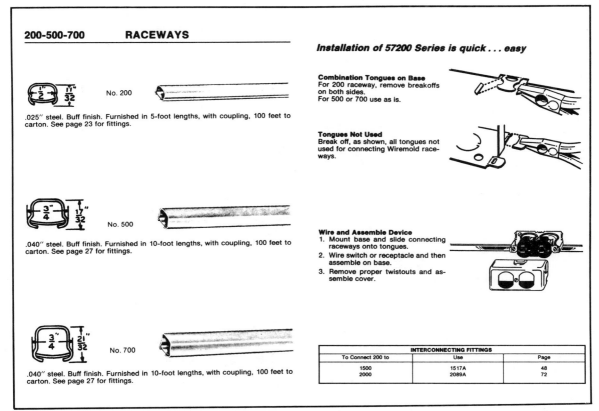

200-500-700   **RACEWAYS**

Installation of 57200 Series is quick . . . easy

No. 200

.025″ steel. Buff finish. Furnished in 5-foot lengths, with coupling, 100 feet to carton. See page 23 for fittings.

Combination Tongues on Base
For 200 raceway, remove breakoffs on both sides.
For 500 or 700 use as is.

No. 500

.040″ steel. Buff finish. Furnished in 10-foot lengths, with coupling, 100 feet to carton. See page 27 for fittings.

Tongues Not Used
Break off, as shown, all tongues not used for connecting Wiremold raceways.

Wire and Assemble Device
1. Mount base and slide connecting raceways onto tongues.
2. Wire switch or receptacle and then assemble on base.
3. Remove proper twistouts and assemble cover.

No. 700

.040″ steel. Buff finish. Furnished in 10-foot lengths, with coupling, 100 feet to carton. See page 27 for fittings.

INTERCONNECTING FITTINGS		
To Connect 200 to	Use	Page
1500	1517A	48
2000	2089A	72

Fig. 13-2. Illustration of Wiremold surface raceway, giving dimensions of series: 200, 500, 700. (Courtesy Wiremold Co.)

Wiremold is installed by attaching the supporting clips to the surface, snapping the Wiremold into the clips. The wires are then pulled in to complete the installation. The supporting clips are mounted using the anchors appropriate for the surface. Electrical wholesalers carry Wiremold, but even at retail prices you will save money by doing it yourself.

WIRE TYPES

Wires are technically known as *conductors* and they are bare or insulated, depending on the usage. Wires used in house wiring are either rubber-covered with a braid (RHW) or thermoplastic-covered (TW or THHN). Sizes commonly used in house wiring range from No. 14 (No. 18 is used only in wiring ceiling and wall fixtures) to Nos. 4, 3, 2, 1, and 0. Lightweight extension cords may have No. 16 wire. The larger the number, the smaller the wire. The numbers refer to the American wire gauge (AWG). See Fig. 13-8.

Each size wire has an allowable ampacity (or maximum) determined to be safe for its size and type of insulation. If this ampacity is exceeded, overheating and a fire could result. Wires No. 6 and larger must be stranded. Solid wire in sizes No. 6 and larger would be impos-

147

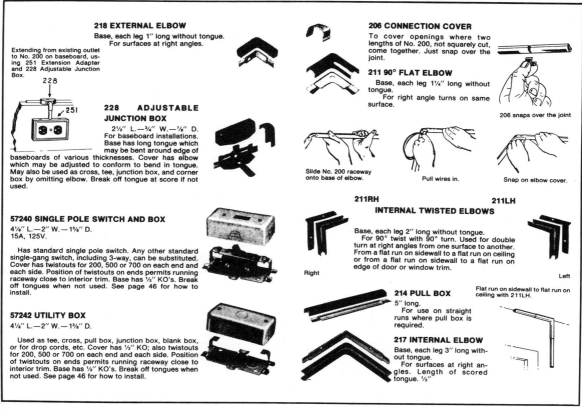

218 EXTERNAL ELBOW
Base, each leg 1″ long without tongue.
For surfaces at right angles.

Extending from existing outlet to No. 200 on baseboard, using 251 Extension Adapter and 228 Adjustable Junction Box.

228

251

228 ADJUSTABLE JUNCTION BOX
2½″ L.—¾″ W.—⅞″ D. For baseboard installations. Base has long tongue which may be bent around edge of baseboards of various thicknesses. Cover has elbow which may be adjusted to conform to bend in tongue. May also be used as cross, tee, junction box, and corner box by omitting elbow. Break off tongue at score if not used.

57240 SINGLE POLE SWITCH AND BOX
4⅛″ L.—2″ W.—1¾″ D.
15A, 125V.

Has standard single pole switch. Any other standard single-gang switch, including 3-way, can be substituted. Cover has twistouts for 200, 500 or 700 on each end and each side. Position of twistouts on ends permits running raceway close to interior trim. Base has ½″ KO's. Break off tongues when not used. See page 46 for how to install.

57242 UTILITY BOX
4⅛″ L.—2″ W.—1¾″ D.

Used as tee, cross, pull box, junction box, blank box, or for drop cords, etc. Cover has ½″ KO; also twistouts for 200, 500 or 700 on each end and each side. Position of twistouts on ends permits running raceway close to interior trim. Base has ½″ KO's. Break off tongues when not used. See page 46 for how to install.

206 CONNECTION COVER
To cover openings where two lengths of No. 200, not squarely cut, come together. Just snap over the joint.

211 90° FLAT ELBOW
Base, each leg 1¼″ long without tongue.
For right angle turns on same surface.

206 snaps over the joint

Slide No. 200 raceway onto base of elbow.

Pull wires in.

Snap on elbow cover.

211RH   211LH
INTERNAL TWISTED ELBOWS
Base, each leg 2″ long without tongue.
For 90° twist with 90° turn. Used for double turn at right angles from one surface to another. From a flat run on sidewall to a flat run on ceiling or from a flat run on sidewall to a flat run on edge of door or window trim.

Right   Left

Flat run on sidewall to ceiling with 211LH.

214 PULL BOX
5″ long.
For use on straight runs where pull box is required.

217 INTERNAL ELBOW
Base, each leg 3″ long without tongue.
For surfaces at right angles. Length of scored tongue: ½″

Fig. 13-3. Representative selection of Wiremold fittings. (Courtesy Wiremold Co.).

sible to handle to pull through conduit and bend into place in electrical panels.

Romex and BX are used in sizes No. 14, 12, 10 to No. 2. Cables can have two or three wires inside, always with the additional grounding wire included. Do not buy or use cable with no ground. It is usually illegal. All cable and wires are marked for use up to 600 volts.

Because Romex and BX are very popular as wiring materials, they are available in any length—even by the foot. I would recommend that you use No. 12 Romex in your wiring jobs, because by so doing you will be buying in full

coils and getting the quantity price. In addition, all the branch circuits can then be protected by 20A circuit breakers.

Wiring with Thinwall is more expensive and it is usually not required. See Fig. 13-9. Before proceeding with any wiring project, contact the governing authority to determine the local electrical requirements before purchasing a lot of materials and then finding that you cannot use them. Most home improvement centers now sell all types of wiring materials by the foot (even single wire used with Thinwall). In using Thinwall or other conduit, it is necessary to know

HOW TO INSTALL WIREMOLD
200, 500 and 700 RACEWAYS

Each length of raceway is furnished with a coupling.

1st. Push out coupling.

2nd. Fasten to surface, using No. 6 flat-head wood screws with No. 200, and No. 8 flat-head wood screws with Nos. 500 and 700.

3rd. Couple lengths together.

4th. Pull wires in.

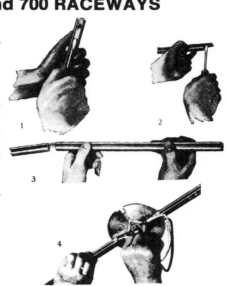

HOW TO INSTALL
WIREMOLD 200, 500 AND 700 FITTINGS

1st. Line up run and fasten base of fitting to surface, using a No. 6 flat-head wood screw for No. 200 and No. 8 flat-head wood screw for Nos. 500 and 700, or the equivalent size expansion shield, toggle bolts, etc.

2nd. Couple base of fitting to raceway by slipping tongue of fitting under the base of the raceway (between the curled edges). When 57200 series fittings are used with 200, break-offs in combination tongues for 200, 500, and 700 should be removed.

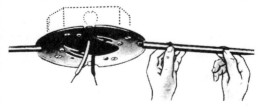

Fig. 13-4. How to install Wiremold raceways and fittings. (Courtesy Wiremold Co.).

200 WIRE CAPACITY	Single Conductor						
TYPE	No. 6	No. 8	No. 10	No. 12	No. 14	No. 16	No. 18
THHN, THWN	—	—	—	3	5	6	7
TW	—	—	—	3	3	4	5
THW	—	—	—	2	2	4	5
RHH, RHW	—	—	—	—	—	4	5

500 WIRE CAPACITY	Single Conductor						
TYPE	No.6	No. 8	No. 10	No. 12	No. 14	No. 16	No. 18
THHN, THWN	—	2	4	7	9	10	12
TW	—	2	3	4	6	7	9
THW	—	—	2	3	4	7	9
RHH, RHW	—	—	2	2	2	7	9

700 WIRE CAPACITY	Single Conductor						
TYPE	No. 6	No. 8	No. 10	No. 12	No. 14	No. 16	No. 18
THHN, THWN	2	3	5	8	11	12	15
TW	—	2	4	6	7	9	11
THW	—	2	3	4	5	9	11
RHH, RHW	—	—	2	2	3	9	11

Note: No. 16 and No. 18 not covered by U.L. Guide Cards.
Capacities are given as a guide to installers of low-potential wiring.

Fig. 13-5. Special fittings for Wiremold also wire capacities for 200, 500, 700 series. (Courtesy Wiremold Co.).

DEVICES: CURRENT-CARRYING ITEMS

In electrical wiring, the word *device* means any component part that carries current, but does not consume it. Sockets, switches (Fig. 13-10), receptacles, and circuit breakers are devices. Power-consuming items are called *utilization equipment* and could be motors, toasters, lamps, etc. These items put a load on the circuit and spin the meter disc.

Wiring devices are many and varied, as you can tell by visiting the electrical supplies counter in a hardware store. Each year brings new devices to the market. Some are good and some are not. Be sure to look for the Underwriters' label (UL) on any device you purchase. This listing label gives assurance of a quality product that will serve the purpose for which it was designed. Voltage and amperage ratings are stamped or printed on the device and its container. In us-

ing any device, keep within these limits or buy a higher-rated device.

OVERCURRENT DEVICES

If a wire is carrying too much current (amperes), it will overheat and damage the insulation. If insulation overheats, it could melt, become brittle, and break or drop off. This leaves a bare wire to cause more trouble. To overcome this dangerous condition, overcurrent devices were developed to limit the current flow in the wire. This maximum depends not only on the wire size but the insulation material and the number of wires permitted in each size of conduit. Excerpts from NEC Tables 3-A, 3-B, and 3-C are shown in the Appendix.

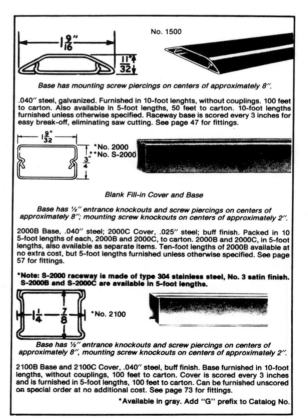

Fig. 13-6. Special Wiremold raceway for overfloor use. Also shows larger square 2000 and 2100 series. (Courtesy Wiremold Co.).

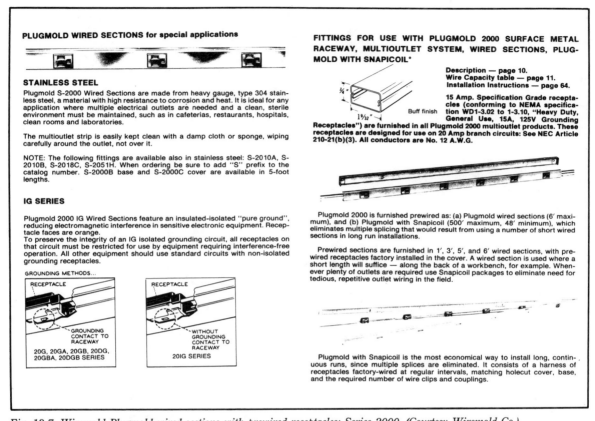

PLUGMOLD WIRED SECTIONS for special applications

STAINLESS STEEL

Plugmold S-2000 Wired Sections are made from heavy gauge, type 304 stainless steel, a material with high resistance to corrosion and heat. It is ideal for any application where multiple electrical outlets are needed and a clean, sterile environment must be maintained, such as in cafeterias, restaurants, hospitals, clean rooms and laboratories.

The multioutlet strip is easily kept clean with a damp cloth or sponge, wiping carefully around the outlet, not over it.

NOTE: The following fittings are available also in stainless steel: S-2010A, S-2010B, S-2018C, S-2051H. When ordering be sure to add ''S'' prefix to the catalog number. S-2000B base and S-2000C cover are available in 5-foot lengths.

IG SERIES

Plugmold 2000 IG Wired Sections feature an insulated-isolated ''pure ground'', reducing electromagnetic interference in sensitive electronic equipment. Receptacle faces are orange.
To preserve the integrity of an IG isolated grounding circuit, all receptacles on that circuit must be restricted for use by equipment requiring interference-free operation. All other equipment should use standard circuits with non-isolated grounding receptacles.

GROUNDING METHODS...

RECEPTACLE

GROUNDING CONTACT TO RACEWAY

20G, 20GA, 20GB, 20DG, 20GBA, 20DGB SERIES

RECEPTACLE

WITHOUT GROUNDING CONTACT TO RACEWAY

20IG SERIES

FITTINGS FOR USE WITH PLUGMOLD 2000 SURFACE METAL RACEWAY, MULTIOUTLET SYSTEM, WIRED SECTIONS, PLUG-MOLD WITH SNAPICOIL*

Description — page 10.
Wire Capacity table — page 11.
Installation Instructions — page 64.

Buff finish

15 Amp. Specification Grade receptacles (conforming to NEMA specification WD1-3.02 to 1-3.10, "Heavy Duty, General Use, 15A, 125V Grounding Receptacles") are furnished in all Plugmold 2000 multioutlet products. These receptacles are designed for use on 20 Amp branch circuits: See NEC Article 210-21(b)(3). All conductors are No. 12 A.W.G.

Plugmold 2000 is furnished prewired as: (a) Plugmold wired sections (6' maximum), and (b) Plugmold with Snapicoil (500' maximum, 48' minimum), which eliminates multiple splicing that would result from using a number of short wired sections in long run installations.

Prewired sections are furnished in 1', 3', 5', and 6' wired sections, with prewired receptacles factory installed in the cover. A wired section is used where a short length will suffice — along the back of a workbench, for example. Whenever plenty of outlets are required use Snapicoil packages to eliminate need for tedious, repetitive outlet wiring in the field.

Plugmold with Snapicoil is the most economical way to install long, continuous runs, since multiple splices are eliminated. It consists of a harness of receptacles factory-wired at regular intervals, matching holecut cover, base, and the required number of wire clips and couplings.

Fig. 13-7. Wiremold Plugmold wired sections with prewired receptacles: Series 2000. (Courtesy Wiremold Co.).

whether the wire is enclosed with other cable or in free air.

I can remember as a youngster holding a candle while my father replaced a fuse in the fuse holder in an upstairs bedroom. The actual fuse was a strip of low-melting-point material (similar to wire solder) connected between two terminal screws. The wiring was exposed along with the meter on the bedroom wall. The service drop entered the house at a point outside, came through the wall, through an open knife switch, into the meter, and then through the ''fuses.'' Even this simple setup provided protection for the house wiring.

Fuses

With the exception of the wire solder fuse, plug fuses are the oldest type of fuse. The plug fuse requires no tools to change, but it is only safe if the outer rim only is touched when handling the fuse. When a new fuse is installed, be sure to tighten it well, because a loose fuse can blow or overheat the holder and damage it.

If you look in the fuse window, you will see the actual fusible link. This link has a narrow center part which is the part that blows or melts. An overload will melt the narrow part and it just disappears. A short circuit actually blows the fuse (explodes it) and the window will be black-

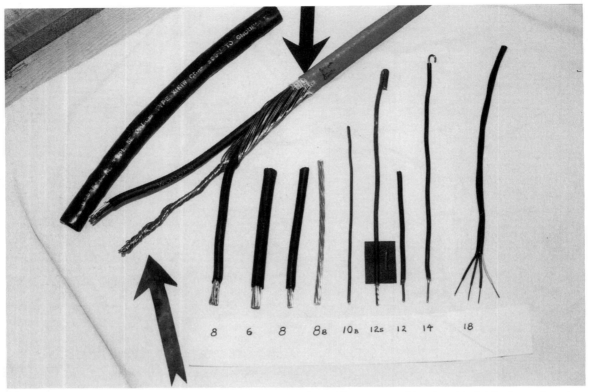

Fig. 13-8. Types of wire. From the left: casing from No. 3 four-wire service-entrance cable, three insulated wires and one bare (used when main breaker is outside of building), standard service-entrance cable, two insulated and one bare, all No. 8, Nos. 6 and 8, insulated, No. 8 bare, No. 10 bare, No. 12 with THHN insulation, Nos. 12 and 14, insulated, and 4-wire No. 18 insulated thermostat cable.

ened. These indications will give some hint as to the type of trouble on this circuit. Plug fuses 15A and smaller have a hexagon window. Larger fuses up to 30A have a round window. Always replace a defective fuse with one of the same size. Plug fuses may be only used on circuits of 150 volts to ground.

Time-Delay Plug Fuses

Time-delay plug fuses, known by the trade name of Fusetron (now called Tron), are designed to carry a greater current for a short period (about one minute) such as is developed

when a motor starts. Examples are washing machines, furnace blower motors, and air conditioners. This current draw could be three to five times the normal running current, although it lasts only a few seconds. This current in-rush would blow a standard fuse. Therefore, the time-delay fuse was developed.

The delay fuse consists of two parts: a *fusible strip* just like the standard fuse and a *solder pot* at the bottom of the fuse body. The solder pot has a wire that is inserted into the solder in the pot. The other end of the wire is attached to a spring. The wire also is connected

Fig. 13-9. An example of a Thinwall offset using ¼-inch Thinwall.

Fig. 13-11. This is a combination dual toggle switch installed in a metal wall box. This device is preferred over the interchangeable type as there is less chance of flashover.

Fig. 13-10. A switch wired into a circuit and ready to be pushed into a wall box. The box is secured to the wall using Madison anchors. The arrow points to the "pound-on" grounding clips. Use care when pounding these clips on the box, as you may force the box and anchors through ⅜- or ½-inch drywall.

to the standard fusible link that is part of the standard fuse type. The spring puts tension on the wire in the solder. When an overload occurs, the solder starts heating. If the overload is of short duration (say the motor starts promptly), the solder does not melt and nothing happens (the fuse does not blow).

If the motor has tight bearings and cannot start within a reasonable time or is unable to turn, the solder melts and releases the wire because the spring pulls it out. If a short circuit occurs, the fusible link blows. Both methods serve the same purpose: to protect the circuit and equipment.

Type S Nontamperable Fuse

The type S fuse has the same characteristics as the Edison-base, time-delay fuse (Fusetron or Tron). The Type S fuse has a smaller threaded part so that a coin cannot be inserted nor can foil be used (threads will chew up the foil). These fuses are called Fustats when made by Bussman Company. They are used in Edison-base fuse receptacles by inserting an adapter. The adapter has a prong on the side that effectively prevents the adapter from being unscrewed from the fuse receptacle.

Fustats and their adapters are sized together so as to prevent overfusing the circuit. This means a 0-to-15-amp adapter will only accept a 0-to-15-amp Fustat. All types of plug fuses may be used on circuits not exceeding 150 volts to ground. This allows them to be used on residential circuits, because with 240-volt service, either side of the 240 volts is only 120 volts to ground. The 240 volts is between the two *hot* wires.

Cartridge Fuses

Cartridge fuses are different in shape from plug fuses. They are cylindrical in shape. Up through 30A they are 9/16 inches in diameter by 2 inches long. The next larger size, 35A through 60A, are 13/16 inches diameter by 3 inches long. The ends are made of brass while the center is made of fiber.

The actual fuse link is inside the fiber center and is surrounded by an inert powder to smother the arc when the fuse blows. Fuses larger than 60A are of the knife blade type. The appearance is similar to the smaller-capacity fuse, but on each end are flat copper blades—hence the knife blade designation.

These blades slide into parallel jaws to make contact with the fused disconnect wiring. Use of the term cartridge comes from the similarity to a rifle cartridge. The 70A size is 1 inch in diameter by 5⅞ inches long. Larger capacities are larger in physical size. The cartridge fuses from 30 to 60 amps are used to protect

ranges, clothes dryers, and water heaters. Modern installations are usually the pull-out-type fuse block.

In a situation where they are exposed, the fuse must be able to be de-energized by a disconnect to make it safe to remove. Even then, use a fiber fuse puller to remove and replace the fuse, which keeps your hand all the further from any live wires or terminals.

The above types are made for circuits up to 250 volts. Other types are for up to 600 volts. These high-voltage types are not used in the home. *Renewable* fuses are made in both the cartridge and knife blade type. Both ends are removable and the fusible link can be replaced. These are usually found in factories and office buildings. The first cost is higher but replacement links are very reasonable.

Circuit Breakers

Circuit breakers, commonly called breakers, are the latest development in overcurrent devices. They can be opened or closed manually and they open automatically on a short circuit or overload. The amount and duration of the overload is determined by the setting of the breaker. Circuit breakers incorporate the time lag feature and carry overloads for a short time, but they trip immediately on a short circuit.

SOCKETS

Various types and sizes of sockets are available for general lighting and decorative effects. Sizes vary from the candelabra at 0.465 inches diameter, intermediate at 0.65 inches, medium (Edison base) at 1.037 inches, to mogul at 1.555 inches. There are other specialized sockets for industrial and commercial use. Three-way table and floor-lamp sockets contain an extra contact for the special three-way light bulbs so that three levels of light are available.

INTERCHANGEABLE DEVICES

Interchangeable devices were developed to accommodate more than one device in a single gang wall or handy box. These separate, smaller devices are:

- ▶ single outlet ground receptacle
- ▶ toggle switch, single pole or three way
- ▶ pilot light
- ▶ push button
- ▶ blank insert

The devices can be combined in any arrangement of the five types listed. A steel mounting strap holds them in position to accommodate a cover plate, similar to a single device in the same space. This strap is furnished with one or three openings. When used for only two devices, the center opening is not used. Devices are inserted from the back of the strap and locked in place by prying a thin portion of the bar over into a groove in the device. Cover plates come with one, two, or three openings. The one-opening strap (and cover) position the device vertically. The two- and three-device positions are horizontal.

Although these devices are listed for all the uses that standard devices are listed for, I prefer not to use them. They take up more box space, especially when three devices are used in a single gang box. Many more wires are introduced into a small wall box and this might exceed the rated capacity of the box. In any event, wiring connections must be made very carefully, and the wires must be folded neatly back in when fastening the assembly to the box ears. Where wall space is at a premium, these

devices are useful. If you are able to use standard devices, do so, because they are easier to work with.

I have found lightning is apt to burn across these devices because of the closeness of the terminals. I found after replacing one switch assembly that the second one also shorted out. I then replaced the assembly with standard switches in separate boxes and had no more trouble.

COMBINATION DEVICES

Combination devices are similar to interchangeable devices. They differ because they are furnished in selected combinations such as: switch/outlet, switch/switch, and switch/pilot light. These combinations are cast in one plastic housing and must be purchased in the arrangement desired. Only two devices are combined in one housing. Standard duplex receptacle plates are used. This type is preferred over the interchangeable style. See Fig. 13-11.

WALL RECEPTACLES

Wall receptacles are as varied as switches. The old types had a screw socket (same as a light socket) with a brass plate and a hinged cover. Lamps and extension cords had a plug with a freely rotating brass threaded shell. When the shell was screwed into any socket, including this wall socket, the cord would not twist. Later, the threaded part of this plug was made separately from the other half which had the two blades. This then became the standard attachment plug we now know. These plug bases are still available.

The first wall receptacle I remember was mounted 7 feet high on my parents' bedroom wall. It was recessed porcelain and had a brass

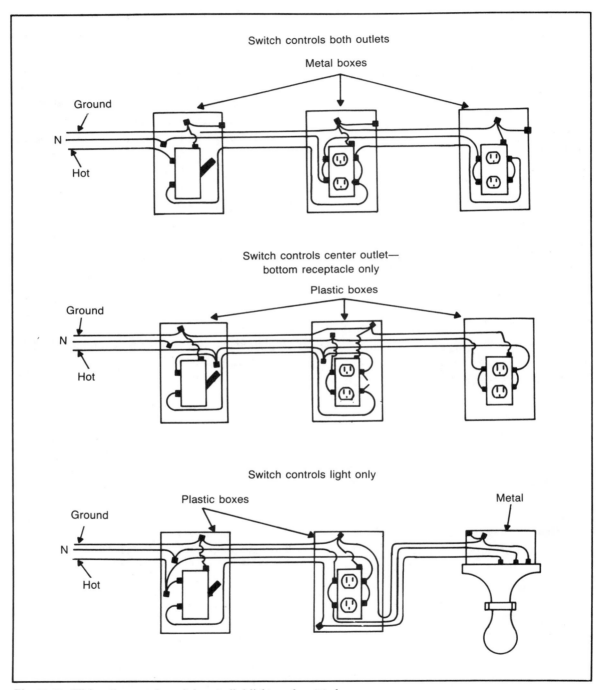

Fig. 13-12. Wiring diagrams for switch-controlled lights and receptacles.

Fig. 13-13. A photocell that controls perimeter lighting.

plate with small hinged doors that opened outward like cupboard doors. Inside was the recess, about 1½ inches high, 1 inch wide, and about 1½ inches deep. On each side was a brass plate (the two contact bars). The plug consisted of a porcelain block with a contact bar on each side that, when inserted into the recess, made contact with the mating bars in the receptacle recess. The two brass doors of the cover plate closed around an extension cord consisting of two twisted wires (green cloth covered) that were attached to the plug. At the other end of the cord was an open socket. This was my first "contact" with electricity. I found out what was in the socket.

Receptacles have different configurations of openings in their faces. One reference shows 18 different receptacles for as many applications. You may be familiar with the difference between electric range and electric dryer receptacles. The range receptacle accepts three straight blades while the dryer receptacle will only accept two straight and one L-shaped blade. This prevents a dryer cord from being plugged into

a higher amperage range receptacle.

Receptacles for lamps, radios, televisions, vacuums, etc., are fused at 15 amps and are the common type found in homes. Inspect one and note the longer slot, usually on the left. This slot is connected to the neutral (white) wire. The newer radios and other appliances now have one wide blade that prevents the plug from being inserted the wrong way. This helps to ground the equipment better.

Some equipment has one wire grounded to the frame and it needs this polarized plug. Receptacles for many years have had one slot wider than the other. This was the design even before the grounding receptacles were developed. The wide slot is connected to the grounded wire (the white wire). When used with a polarized plug, this effectively grounds the frame of the appliance thus connected.

The configuration of 20A receptacles provides one T-shaped slot, usually on the left, as 20A plugs have one prong at right angles to the other. The other slot is vertical (as shown in Fig. 8-4). The T-shaped slot will also accept the standard 15A attachment plug. Receptacles in the kitchen and dining room are on 20A circuits. Generally only 15A receptacles are installed on these circuits as standard 15A plugs are furnished with appliances used in these areas. This is an NEC-approved method.

TIMERS, PHOTOCELLS, AND CONTACTORS

Timers control many current-using items such as lights, thermostats, water heaters, pumps, and so on. They are in series with current-using items and are actually automatically operated switches. Photocells are also switches, but they are dependent on light levels in their area. They are completely automatic and

Fig. 13-14. A wall box held in place using Madison anchors. The two upper wings have been bent inside the box and squeezed against the box wall. The bottom wings have been partially bent in.

need no adjustment. Timers do need to be adjusted for changing sunset and sunrise hours while photocells do not. They react to sunlight. Buy a good brand. I installed one and later the neighbor complained that it blinked on and off all the time. I purchased a better one and had no more trouble. See Fig. 13-14.

Contactors are switches with the capacity to carry larger sized loads. They are operated by a magnetic coil energized by either a photocell, timer, or other controller. Parking lot, stadium, or other heavy lighting loads are operated by contactors. Your home central air conditioner has such a contactor built into the control system.

Ground Clamp

This device clamps tightly to a water pipe or ground rod by means of two screws clamp-

ing the two parts. The grounding wire is held by a setscrew, making a tight connection mechanically *and* electrically.

Electric Utility Meter

Even though this is utility property, it is still a device because it carries current. It is also current using as it uses a small amount to "spin its wheels" so you can be billed. If you must remove the meter from its base or open the meter cabinet, notify the utility. If you remove the meter to work on the service-entrance panel or other hot wires, be sure to put a circle of corrugated cardboard in place of the meter. Be cautious! Notifying the utility beforehand and afterward shows that you are not being dishonest.

WIRE HOLDING ITEMS

Wire holding (containing) items are numerous. See Figs. 13-15 through 13-22. Conduits have been covered in Chapter 7. Other items are, cabinets, pull boxes, and wall and ceiling boxes. Long ago, boxes were not used. Wires were run inside walls using *loom*, a formed, fiber covering shaped like tubing. The wire was run through this loom from the last porcelain knob in the attic or cellar to the "turn" switch on the wall. There were no wall receptacles. It then was decided that because of fire and safety hazards, metal boxes should be used wherever splices or connections to terminals were located. This has now become mandatory.

Boxes must be supported in walls and ceilings so that they will be mechanically solid and secure. Ceiling boxes must be anchored so they can support heavy chandeliers that sometimes weigh 20 pounds or more. Wall boxes are supported in new work by nailing them to the side of the stud. Plastic wall boxes have nails placed through ears on the top and bottom of the box.

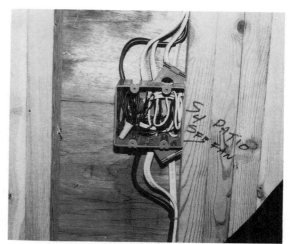

Fig. 13-17. Plastic 2-gang switch box with designations shown. Rough-in is completed.

Fig. 13-15. The reverse side of the wall box installation, showing the opposite side of the Madison anchors.

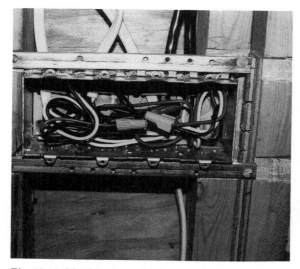

Fig. 13-16. Metal 4-gang switch box near front door. Controls inside and outside lights. Rough-in has been completed.

Fig. 13-18. Doorbell transformer installed on a blank plate. Voltage of 12V to 18V allows wires to be installed without boxes or protection.

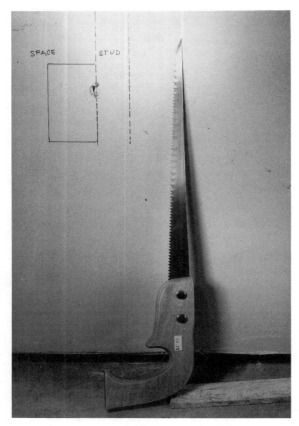

Fig. 13-19. Free space determined for a wall box. The keyhole saw is fine for this work.

Ceiling boxes and wall boxes for fixtures are mounted on adjustable bars that are nailed between studs for wall boxes and between joists for ceiling boxes. These adjustable bars have bent-over ends to be nailed to the framing. The box is mounted in the center, but by loosening the setscrew, the box can be moved so as to center it on the ceiling or to center a wall box on a wall space. Certain wall boxes have removable sides and can be attached together (ganged) to make long horizontal switch boxes for uses such as stores and office buildings. The most a homeowner would need might be a four-gang

for four switches. See Figs. 13-23 through 13-27.

Octagonal boxes resemble a 4-inch square box, but with the corners cut off (see Figs. 13-28 through 13-30). The square, 4-inch box can be used for ceiling fixtures with the addition of a plaster ring. This is a raised cover with a round center ring with ears having tapped holes to accept a fixture bar. This bar comes with every fixture and is part of the mounting kit. Most boxes come with BX or Romex clamps (except the 4-inch square and the larger $4\frac{11}{16}$-inch box; these are used for commercial work). Very shallow ceiling boxes are made one-half inch deep for mounting where a joist is in the way when centering a ceiling box.

The NEC limits the number of wires allowed in any wall, ceiling, handy box, or junction box. This eliminates crowding and the sharp bending of wires. Any box packed full of wires and devices is sure to generate heat. The heat could build up and damage the insulation or the device. Plastic boxes have the limit of the number of wires embossed inside, along with the cubic inch figure. A typical single-gang plastic wall box might have the following designation: "18 cu. inch. - $\frac{9}{14}$, $\frac{8}{12}$, $\frac{7}{10}$." The maximum number of conductors that may be installed in a designated-size box is given in NEC Table 370-6(a).

The extensive use of cabinets in home wiring is unlikely except when used as fuse or breaker cabinets or disconnect cabinets. When installing new service-entrance equipment, a junction box/cabinet must be installed if the new equipment has to be moved (thus leaving the old wires too short to reach). In all boxes of any kind, unused open knockouts must be closed with knockout plugs (snap-in metal plates that seal the opening). Metal plugs in plastic boxes must be recessed $\frac{1}{4}$ of an inch from the outer

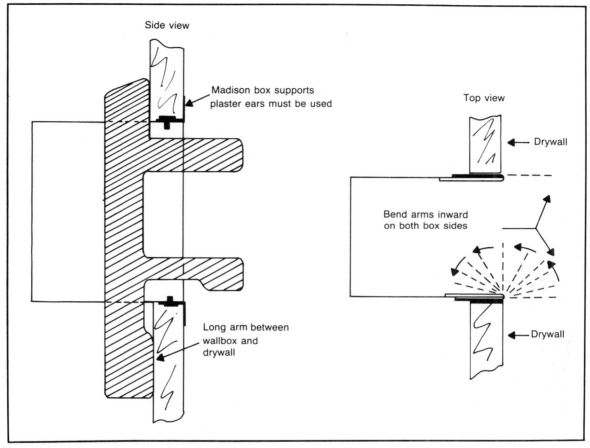

Side view

Madison box supports
plaster ears must be used

Top view

← Drywall

Bend arms inward
on both box sides

Long arm between
wallbox and
drywall

← Drywall

Fig. 13-20. Madison patented "old work" wall box support for metal or plastic boxes. Wall box and two supports are inserted into opening and positioned as shown. Box ears are pushed against drywall and supports are pulled against back side of drywall. Short arms are bent tight against each side wall of box. Use pliers to pinch them tightly.

surface, as stated in code Section 370-8. Try to avoid making extra openings in plastic boxes. It will be hard to close those not used.

Metal raceways are not used in home wiring, but they are common in large installations such as office buildings.

NEW WORK

New work is electrical wiring installed before any of the finish materials such as drywall or paneling are applied to the studs or joists. It also means any place where the framing members are exposed such as basement and attic areas. The term "new work" refers to the method used to install such wiring, not the age of the building. In fact, "old work" can and is installed in *new* buildings where changes and additions to the original construction plans must be made after the building is completed.

Romex and BX in new work are run through holes bored in joists and studs. Cables

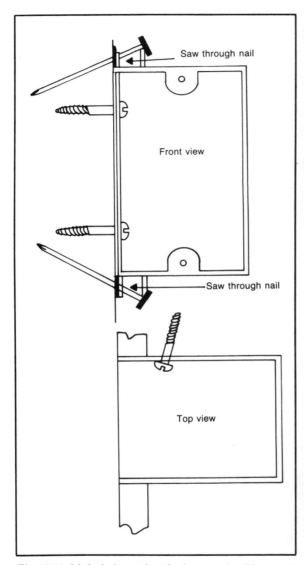

Fig. 13-21. Method of removing plastic or metal wall box to add or change wiring, then replacing box. Put electrical tape over screw heads when using a plastic box.

Fig. 13-22. A ceiling receptacle. The box and plate are selected for future ceiling installation to give proper appearance.

Nailing supports are an integral part of the box construction. On nonmetallic boxes, tight hooks hold nails on the top and bottom. Some metal boxes have nails going through near the top and bottom, inside the box. These come out on the other side to go into the wood stud. NEC Section 370-13 describes this in detail. Other boxes have an angle bracket on one side for nails

Fig. 13-23. Ceiling junction box. The NEC states that all wire connections must be made in enclosed metal or plastic boxes, and covers must be installed on these boxes.

must be anchored within 12 inches of the box and every 4½ feet when run parallel to framing members. Nonmetallic boxes require stapling within 8 inches of the box.

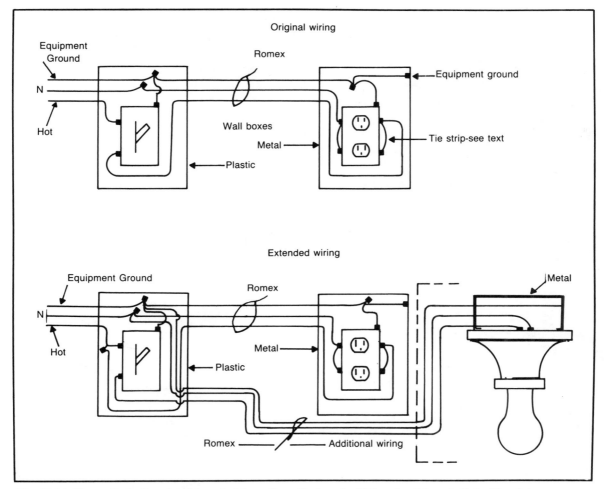

Fig. 13-24. Shows elimination of switch control for wall receptacle and addition of switch-controlled ceiling fixture.

or self-formed nailing prongs. Supports will position the box so that it will be flush with the finished wall surface.

THE WIRING LAYOUT

You must have a blueprint or drawing of a new wiring layout for a new dwelling or an addition. See Fig. 13-31. This might even be required when you apply for a permit. Contact the utility to find out their preferred location for the service-entrance mast and meter. They will advise the best location after conferring with you. Certain locations are preferred; others might be acceptable to the utility. Underground service is not so flexible as to location. Again, the utility has the final say.

If the service is completely underground, you have to furnish conduit from the bottom of the meter housing down into the ground. Some utilities require a 90-degree elbow turned to-

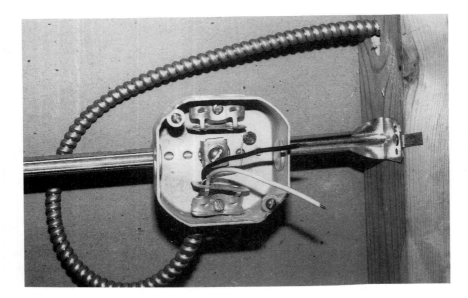

Fig. 13-25. Close-up of ceiling box in Fig. 13-29. Nails and clips keep hanger bar in position flush with the finished ceiling. Grounding wire is secured under screw in box bottom,

ward the supply source. The service might only be underground to the utility pole. Conduit is also required from underground up the pole to a minimum height of 8 feet (or as the utility requires). When talking to the utility representative, mention the proposed location of the interior service-entrance panel. They might suggest an alternate meter location. The overhead service drop restricts the meter location because of clearance requirements. Roof clearance

Fig. 13-26. Drywall in place on a mock-up ceiling. Enlarged opening must be sealed as required by the NEC.

Fig. 13-27. Fixture in place. Note double screw holes to fit an octagonal 4-inch box or a plaster ring.

With the equipment location decided, the branch circuit wiring can now be run. Mark locations for receptacles, switches, ceiling fixtures, and any other junction box locations on the framing and on your print. Mount these items to studs or joists. Next, bore holes for the Romex or BX cables. Route these cables by the most direct path, but do not skimp on length. A little extra on each end of the run is

Fig. 13-28. A 4-inch octagonal ceiling box having BX clamps. Slots in clamp show red bushing (required when using BX).

is 8 feet minimum; ground clearance is 12 feet for 240-volt service (less than 300 volts to ground). Connection at the building requires a minimum clearance of 10 feet (more if possible). The higher the better.

SERVICE EQUIPMENT AND DISTRIBUTION PANEL

Service-entrance/distribution panel equipment must be adjacent to the meter location. Warm climate areas have these items outside. More often, the equipment is inside directly opposite the meter outside. The service-equipment and distribution panel come in one enclosure for the average dwelling. The capacity is 100A minimum, often 150A. See Fig. 13-32.

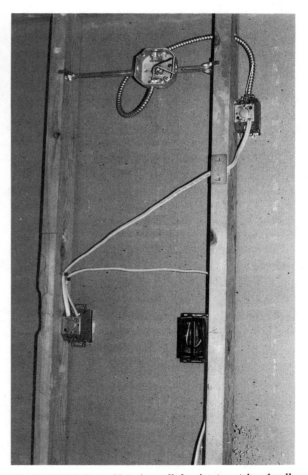

Fig. 13-29. Mock-up of interior wall showing two styles of wall boxes and an octagonal ceiling box mounted on an adjustable bar hanger. Notice steel plate over notch in stud to protect Romex cable.

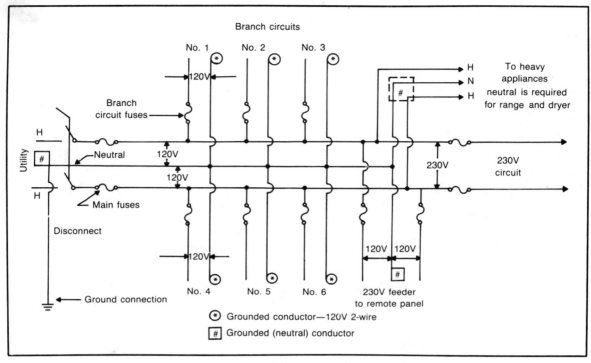

Fig. 13-30. Wiring layout for a residence. This is the standard layout for the modern residence.

better than a length 6 inches too short. True, a short length can be pulled out and used elsewhere, but that takes time and labor.

The NEC requires that no wall space be more than 6 feet from a receptacle. Also, any wall space 2 feet or wider must have a receptacle. Counter tops wider than 12 inches must have a receptacle. Bathrooms, outdoors, and laundries must have one receptacle each. Boxes for switches are 46 inches high. Boxes for receptacles are 15 inches high.

BRANCH CIRCUITS

With the location of devices established, plan the layout of the branch circuits. For residential occupancies, the code requires 3 watts per square foot of floor area (measuring the outside dimensions of the building). Include a basement for a possible recreation room and workshop. Not less than one circuit should be installed for each 500 square feet of area. Plan to have most areas served by two separate circuits; if one goes out, the other will still be usable. These circuits can be 14-2 Romex with ground and be protected by a 15A fuse or breaker. These circuits could also be 12-2 with ground protected by 20A breaker or fuse.

Small Appliance Branch Circuits

Outlets in the kitchen, pantry, family room, dining room and breakfast room must be supplied by two or more 20A appliance circuits wired with 12-2 with ground Romex. No other outlets shall be on these circuits. The kitchen

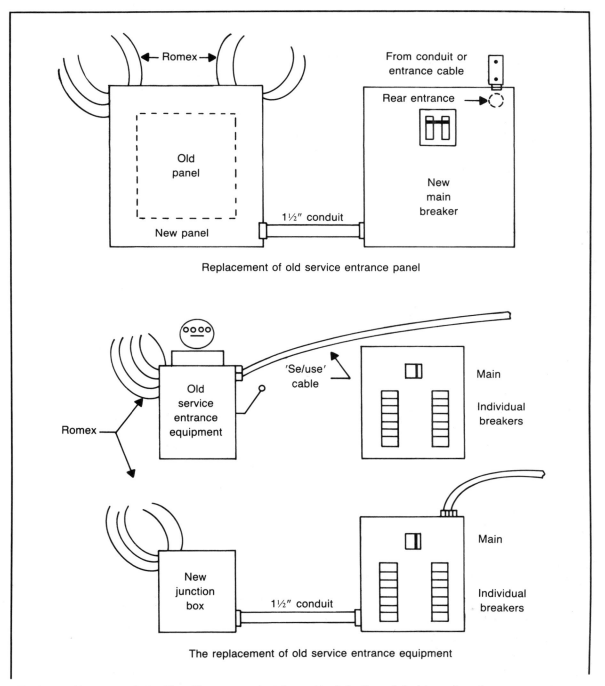

Fig. 13-31. Alternate methods of installing new service-entrance panels in place of obsolete equipment.

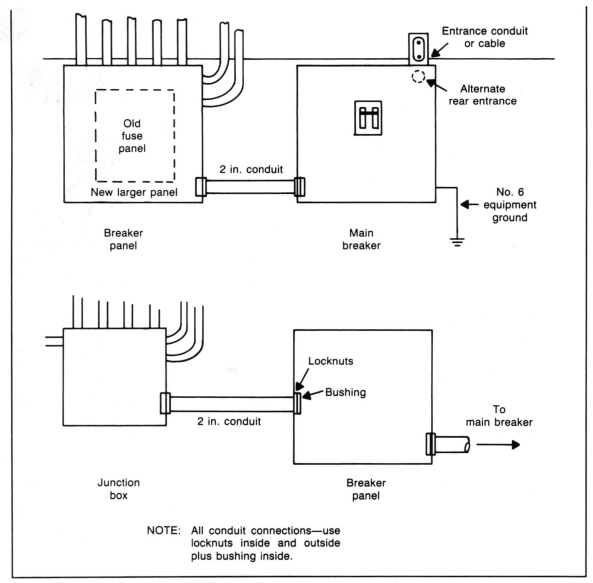

Fig. 13-31A.

area must have outlets supplied by two circuits. This way, more than one high wattage appliance can be used at one time without blowing a fuse. Laundry outlets must be on a separate 20A circuit.

Electric Ranges

The feeder loads for electric ranges or other cooking appliances are calculated from NEC Table 220-19. Range circuits are sized according to a demand factor rather than the ac-

Fig. 13-32. Close-up range outlet rough-in. Cable has four wires: black, white, red, and bare. In situations where range feed does not originate at the service-entrance panel, three wires and the bare equipment ground are required.

tual nameplate rating (connected load). This is because all heating elements of a range are seldom on at one time. If the cooking units consist of one cook top and not more than two wall-mounted ovens, it is permitted to add the nameplate ratings together and treat this as one range (provided all components are in the same room). Note 1 to Code Table 220-19 explains how to arrive at the demand rating if there is more than 12 kW for one range. Carefully read all notes to tables as you use them. See Fig. 13-33.

Electric Clothes Dryer

The electric clothes dryer demand factor is 100 percent of the nameplate rating. It is recommended that No. 8 wire be used for this circuit and be protected with a 40A breaker. No. 10 wire can be used, protected by a 30A breaker, but the heating elements in clothes dryers are being made in heavier wattages. When a dryer replacement is needed, the circuit might not handle it and will need to be rewired.

Heating and Cooling Loads

Because heating and cooling equipment is never used simultaneously, the effective load is taken as the cooling load. This load is the greater of the two loads. This load is taken at 100 percent of the nameplate rating. As an example, my own air conditioner load is 19.0 amps while the heating load is 2.4 amps. Therefore the effective heating/cooling load is included as 19.0 amps × 240 volts = 4560W.

OLD WORK

Old work refers to additional wiring installed in those buildings already finished. It also refers to repairs or the replacement of obsolete wiring. This work is much more difficult than new work because of the problems encountered in fishing cables through walls and ceilings. All wiring is concealed for appearance and protection.

A one-story dwelling is fairly easy to rewire because the attic and basement or crawl space should be accessible for work. The rewiring work can be done one section at a time or one circuit at a time. Not only should an old dwelling be rewired, but it should be updated and wired for today's needs. Allow for additional circuits not included in the old wiring system. Install three-way switches so that you can walk through the house without retracing your steps to turn lights on and off. In new dwellings, the outlets in the various rooms—even though they are located according to NEC requirements— are sometimes located at inconvenient spots.

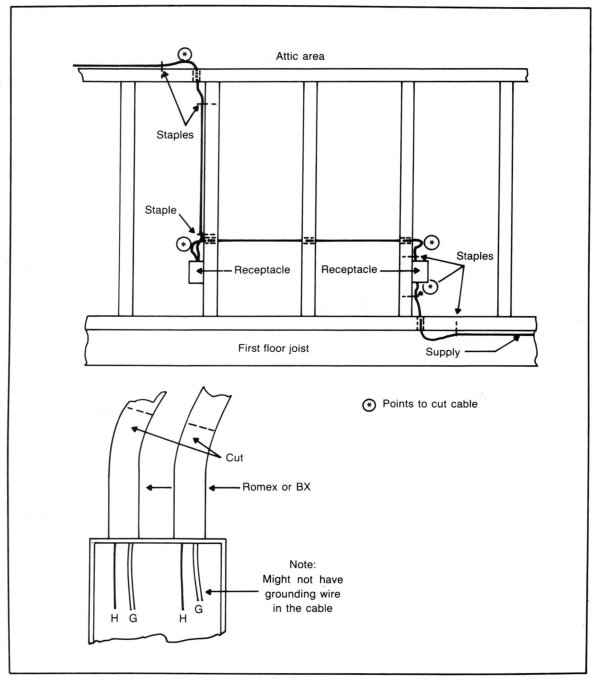

Fig. 13-33. How to disconnect abandoned wiring. Cut back far enough to prevent any possible contact with new wiring.

When doing rewiring, you now are able to locate the outlets where they are most convenient. Just remember to maintain minimum NEC requirements.

REPLACING THE OLD WIRING

Removing and rewiring work can tax your ingenuity a great deal. Some old wiring has to be cut off as far as possible inside walls and pushed out of the way. There can be no possibility of the abandoned wiring contacting any of the new wiring. Reach in through where the old wires came out of the wall and cut them inside the wall itself. If Romex is used, it will be stapled close to the old outlet location so that it cannot be pulled out, so cut it off. See Fig. 13-33.

Other wires might be routed horizontally through the studs and also are stapled. New cable will have to be run either up and over the attic or down and under the first floor (if you

Fig. 13-34. Hole bored through first floor plate for cable from the basement.

are lucky enough to be working on a one-story dwelling). (Refer to Fig. 9-1).

A two-story dwelling can give you other problems. Sometimes a pipe chase (opening through the top and bottom of a wall for pipes) can be used for cable. If you use a pipe chase, check that there are no heating pipes or ducts or even hot water pipes. This would increase the wire temperature and affect the insulation. Many times, access can be gained in closets and other hidden areas, and cutting floor and ceiling openings will not affect the appearance of a regular room or need extensive redecorating.

When running concealed wiring, be resourceful and plan ahead. You might need to run three or more cables up from the basement to supply the second floor circuits. If you have to remove flooring, refer to Fig. 9-19. To "fish" cable up or down finished walls, you need a chalk line at least 20 feet long or longer. You also need a handmade mouse and a handmade "fish hook." See Figs. 13-35 and 13-36.

A mouse is made from 3 inches of ⅛-inch wire solder. The solder is bent back upon itself about 1 inch from one end and then a loop is formed and the rest is wound tightly around the 1-inch end so as to form a compact weight about 1¼ inches long and ½ inch in diameter.

The fish hook is made from coat hanger wire. Use a piece about 12 inches long. One end is bent with a hook in an L shape. The other end is formed into a handle for ease in holding. Make a hand-size loop. The center is formed as needed to reach inside a wall opening and retrieve the mouse hanging from the chalk line inside the wall. (Refer to Figs. 9-1, 9-2, and 9-12 through 9-16.) A lead mouse is best because it is heavy for its size. A large nut could be used, but it might not go through holes bored in the framing or wall.

To find studs in a finished wall, use a commercial stud finder or tap with a screwdriver or your knuckles. A long, slim nail pushed through drywall or plaster can locate studs if they are hard to locate otherwise. Or, make a stud finder from a coat hanger. Form a loop handle and bend an angle on the other end. Push this through a nail hole in the drywall and rotate the bent end. Align the loop handle with the bent end so as to know which way the bent end is pointing. These small nail holes are easy to patch. Taking off the baseboard sometimes helps you locate studs. Check carefully before cutting a large opening for a wall box. Such holes are hard to patch. See Figs. 9-2 and 9-12.

When working alone, you will be running back and forth, dropping the mouse inside the

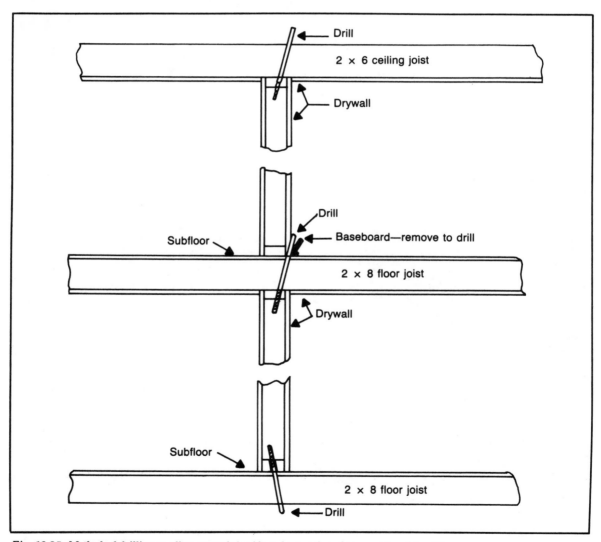

Fig. 13-35. Method of drilling to allow concealed cable to be run from basement panel to attic.

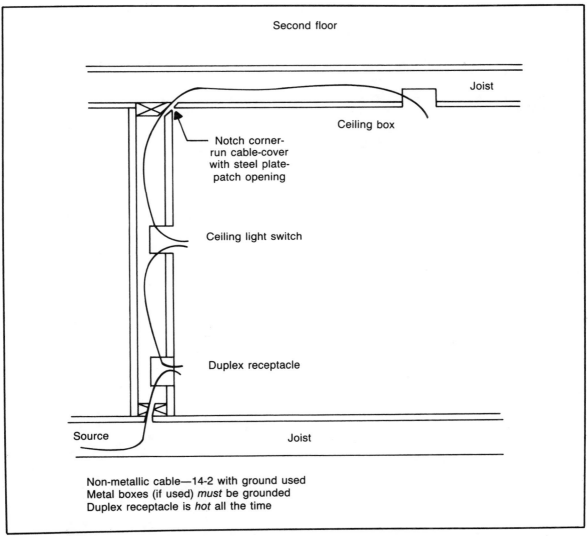

Second floor

Joist

Ceiling box

Notch corner-
run cable-cover
with steel plate-
patch opening

Ceiling light switch

Duplex receptacle

Source

Joist

Non-metallic cable—14-2 with ground used
Metal boxes (if used) *must* be grounded
Duplex receptacle is *hot* all the time

Fig. 13-36. Method of running cable to supply a switch, ceiling fixture, and receptacle. Note methods of access to various wall cavities.

wall and fishing for it from below. Be sure to have enough length of cord for the mouse to hit the bottom of the wall inside the partition. When the mouse hits the bottom, raise it slightly so that the cord hangs free and does not lie against either inside surface of the hollow partition. Hold the mouse just free of the bottom by laying a weight on the cord at the upper end, but do not tie it. If you hook the mouse, you can then pull it out from inside the wall without any trouble. If you tied the top end of the cord, you would not be able to pull out the mouse from the inside. At a greater distance back on the cord, tie it to something so the cord does not

Fig. 13-37. A two-gang plastic wall box, not roughed in. In new work, boxes are mounted and information is printed on the framing next to the box as to what wiring is to be run to it.

saw blade (less frame) between the box and stud. Saw both nails completely through and the box should now be loose from the stud. When you use a hacksaw blade, put tape on the end you hold to prevent injury. Pull the box from the wall and remove the cables. You can now add additional cables or reroute those already there, depending on what you want to control.

When you are ready to replace the box, drill two holes in the side of it to pass a No. 10 round-head wood screw. Drill one hole ½ inch down from the box top and 1 inch in from the front edge. Drill another hole ½ inch up from the box bottom and 1 inch in from the front edge. Now replace the box in the opening with the cables in place. Hold the box flush with the finished wall and drive two screws in the holes you just

fall completely inside the wall. This procedure seems complicated but is very simple in actual practice. Remember to double-check your measurements, because a 4-inch error can bring the hole in the ceiling of the room rather than inside the wall. It is best to work carefully and not make mistakes.

When you want only to modernize or extend your present wiring, this is "old work." Such a situation would be where a wall switch controls a wall outlet only. You might want to control a newly installed ceiling fixture from the present wall switch and leave the outlet hot all the time. Another way would be to make half of the outlet hot and the other half switch controlled. Many outlets are wired this way where there is no ceiling fixture in the room.

To remove a plastic switch box from the wall, remove the device from the box. The box is usually nailed to the stud next to it. With a screwdriver, gently pry the box—side, top and bottom—from the stud enough to insert a hack-

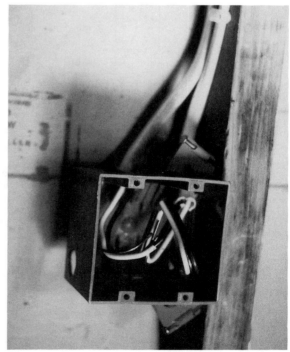

Fig. 13-38. Two-gang plastic wall box. Rough-in has been completed. Hole in left side will be covered before wall finish is installed. (Refer to Fig. 13-37 also.)

made in the box directly into the side of the stud. It is advisable to ground these screws just as if they were part of a metal box. See Figs. 9-21 and 13-21.

You can now connect the new wiring as you planned. Any openings in the boxes not used must be closed, so don't open any more than you will be using. Openings in plastic boxes are hard to close. This original plastic box may be replaced by a metal box and the screws holding the box to the stud would not have to be grounded. Be sure to patch the wall surrounding the box after replacing it in the opening. The metal box, if used, will probably be smaller than the plastic box and will need patching. The NEC requires this.

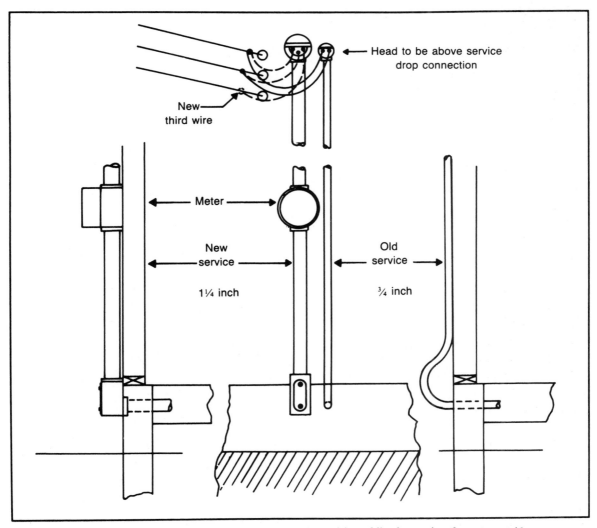

Fig. 13-39. A replacement of obsolete service with adequate modern wiring while also moving the meter outside.

Fig. 13-40. *Slice of a tap onto a line wire. Ready for soldering.*

Fig. 13-41. *A pigtail, soldered. Notice how the solder has flowed into the joint making it mechanically and electrically secure.*

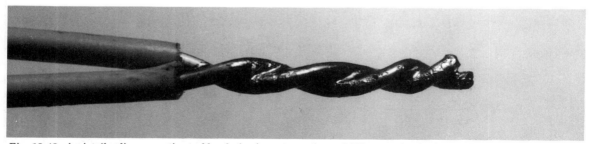

Fig. 13-42. *A pigtail splice connection, soldered. A wire nut may be used if bare section is shortened.*

COPPER SOLDERING TECHNIQUES

The soldering of copper splices has almost become a lost art because of the many devices for connecting wires to each other and to ter-minals. Sometimes wires do have to be sol-dered. As in every soldering job, the material has to be heated and the solder then applied to the hot area (now referring to temperature). If the material or wire is heated high enough, the

Kester "44" Resin Core Solder

Kester's "44" Resin Core Solder—a revolutionary solder advancement—featuring an amazing, fast-acting, instant-wetting, non-corrosive and non-conductive flux that (1) makes soldering faster than ever before and (2) provides a stronger longer-lasting bond.

   Diameters: 1/32" (.031"); 1/16" (.062"); and 3/32" (.093"). Others available.
   Alloys: 40/60, 50/50, 60/40 and 63/37. Others available.
   Available in 1 lb., 5 lb. and 20 lb. spools.

Fig. 13-43. Kester resin-core solder. Available in various diameters, alloys, and quantities. (Courtesy Kester Solder.)

Kester TV-Radio Solder

New and improved product. The small package of Kester Core Solder in the practical (round) wire-size of .062" (1/16"). Ideal for the Serviceman's Kit ... or for those sales to the occasional user for soldering TV, Radio components ... everything electrical or electronic.

Fig. 13-44. Kester TV-radio solder, available in small handyman rolls of 1-ounce each. (Courtesy Kester Solder.)

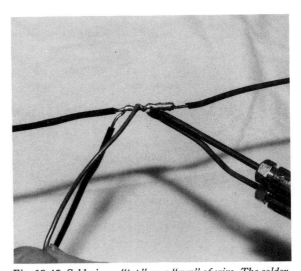

Fig. 13-45. Soldering a "tap" on a "run" of wire. The soldering iron must heat the splice itself, not the solder. The solder is melted and blends with the splice solely from the heat of the splice, producing a perfectly soldered joint with shiny, smooth, evenly blended turns.

solder melts and blends in with the material (copper in this case). To prepare the joint, make sure the wires are absolutely clean. Sometimes sandpaper works; usually a pocketknife is enough. Twist the wires together tightly because the NEC requires that the joint be mechanically and electrically secure *without* solder, and then soldered. All splices and any other exposed current carrying parts must be covered with an insulation equivalent to that used on the conductors themselves. Modern insulation, when removed from the wire, leaves a very clean surface. See Figs. 13-41 through 13-47.

   I remember scraping and scraping the old fabric and rubber insulation from the wires before soldering them (there were no wire nuts then). Every joint had to be soldered. After soldering, the joint had to be taped with rubber tape, then the warm hand tightly gripped the tape to somewhat "vulcanize" it (weld it together, so to speak). Friction tape was added to hold the rubber tape in place.

Tape application methods were as they are now. Start at one end a half inch or so onto the insulated wire. For a "pigtail" splice, continue past the soldered end for about ¾ inch. Fold this flap back to make a thicker end cover, and then continue back toward the starting point. This method actually gives four layers of tape.

The new plastic electrical tape is a great improvement over the old two-tape method. It is faster and it sticks better. The disadvantage, discovered shortly after the plastic tape was introduced, was its extreme thinness, 8.5 mills

Fig. 13-47. Shown is a toggle switch having the terminals covered with electrical tape to avoid contact with the metal box.

(trash bags are 2 mills). Electricians not familiar with this property often did not tape the soldered joints with enough layers on the end of the pigtail joint. Sometimes the tape would be scraped off this end, causing a short circuit. This condition is easily overcome by simply using care when taping. The old rubber tape was much thicker and did not cause this problem. When pushing a plastic-taped joint back into a box, keep the end of the joint away from the box wall (see Fig. 13-47).

This discussion about soldering and taping should be reviewed again when you need to solder electrical wires; just tape your soldered connection with care.

Note: Many types of plastic wire insulation can melt and leave the wire bare at that point. Romex and other types of cable are also prone to excessive heat (above the rating of the insulation, usually a maximum of 90° C). Therefore, use care to keep open flames and other high heat away from electrical wiring.

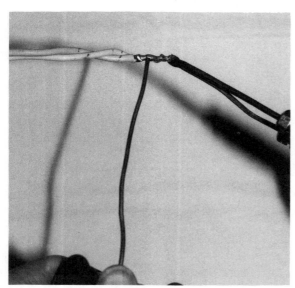

Fig. 13-46. Soldering a pigtail joint, also a perfectly blended joint. A propane torch may be used, but as very little heat is needed, use the smallest tip available. CAUTION: After using a torch, stay around or check back often for at least a half-hour. A spark can ignite something that might smoulder and later burst into flame.

Chapter 14

Special Projects

There are many improvements and additions that can increase the value of a home. A homeowner can save on installation costs by doing the work himself and can save on heating and air conditioning operating costs.

NEW THERMOSTAT

The standard heating thermostat (*stat* in industry terminology) that sells for about $20 can replace your present stat. It has the potential to save you money two ways. You can install it yourself and you can buy it at a discount. In addition, a new stat will operate the furnace more efficiently—saving fuel.

To get the most efficient operation from your equipment, the stat must be properly located. Evaluate the location of the old stat. Living room or family room locations are best. Hallways with good air circulation are the next best locations. If the stat is in a poor location, furnace operation will be affected and may cause

longer or erratic furnace operation; this causes needlessly higher fuel bills. Cold drafts cause the same type of problem. This is why a stat must never be installed on an outside wall. For even more fuel economy, you might want to buy and install a night set-back type stat.

The best stat on the market is the Honeywell Model T87-F. See Figs. 14-1 through 14-4. This three-wire stat can be used for heating only, cooling only, or both heating and cooling, depending on which two wires are used for each function. For both functions, three wires are used. So as not to confuse you, some old stats also had three wires (red, white, and blue). Thermostat cable is color coded because this control work, as it is called, is very dependent on correct wiring, as is line voltage house wiring.

The old three-wire stat was for heating only and it is now obsolete, but it is still in use in many older homes. This stat is Series 10 (the

Fig. 14-1. Honeywell T87-R-28 heating/cooling thermostat. (Courtesy Honeywell Inc.).

T87-F is Series 80). If the Series 10 stat is controlling a relay that is operating a gas valve, oil burner, or hot water circulating pump, the relay will have three terminal screws: red (R), blue (B), and white (W).

To use the T87-F, the R and B terminals of the relay must have a jumper wire connected between them. This can be a short bare piece of No. 18 stat wire. The old stats were the vertical open contact type Series 10. These stats are 4 inches high. The new T87-F, which is round, will not cover the area left by the old stat. Solutions vary from painting the bare spot or papering the bare area. A (very unattractive) cover plate is available from Honeywell. It is about 5 inches in diameter.

To proceed with the installation, first buy the stat. Try to find one in a sealed carton rather than one that has been opened and examined. Some small parts might have dropped out. Mounting screws and instructions are the loose parts. At home, shut off the furnace switch; this is usually on the side of the furnace cabinet. This switch should be plainly marked "main furnace shut-off switch" so that it can be turned off in an emergency.

After you have turned the switch off, you can remove the old stat. The vertical model has one cover screw (top or bottom). Loosen this screw and take off the cover. The three terminal screws will be visible. Loosen them, straighten the wires away from the screws, and

thermostats—single-stage

T87F THERMOSTAT—THE ROUND

PROVIDES TEM-PERATURE CONTROL FOR 24 TO 30 VOLT RESIDENTIAL HEATING, COOLING, OR HEAT-ING-COOLING SYS-TEMS.

Features dustproof, spdt mercury switch, ad-justable heat anticipator, and a 2- or 3-wire wallplate for heating only, cooling only, or heat-ing-cooling systems with remote switching. Add a Q539 subbase in systems requiring system and fan switching at the thermostat location. Q539 also provides cooling anticipator. For T87F mod-els designed to meet Department of Defense specifications, see page 111. Also see Thermostat Guards section, page 96, for THE ROUND WITH KEY LOCK COVER.

APPROXIMATE DIMENSIONS:

	Diameter		Depth	
	in.	mm	in.	mm
T87F	3- 1/4	82.6	1-1/2	38.1
T87F with 137421 Wallplate or Q539 Subbase	3-11/16	93.7	1-3/4	44.5

ELECTRICAL RATINGS:
Mercury Switch: Full Load—1.5 A; Locked Rotor—3.5 A at 30 Vac.
Adjustable Heat Anticipator: 0.1 A to 1.2 A.
Cooling Anticipator: 0 A to 1.5 A, 24 to 30 Vac.

REPLACEMENT PARTS:
104456B Wallplate Assembly, 2 terminals (heating only models). Includes terminal screws.

114854 Crystal, 1.9 in. [48.4 mm] dia.

114855-01370 Thermostat Cover Ring.

ACCESSORIES:
104994A Calibration Wrench.

129044A Adapter Plate Assembly. Includes 6 in. [152.4 mm] cover ring and adapter plate. For mounting T87F thermostat on outlet box and for covering old thermostat mounting marks.

137421A Wallplate for heating and cooling sys-tems. For T87F without positive OFF. Includes spdt heating only (series 20) alternate terminal markings and cooling (also series 20) antici-pator.

137421B Wallplate for heating and cooling sys-tems. For T87F with positive OFF.

TG503A1000 Metal Thermostat Guard. Includes locking cover, backplate, and bracket for mounting on a standard size outlet box. Also includes 104456B 2-terminal Wallplate. Not for use with T87F mounted on 137421 Wallplate or Q539 Thermostat Subbase.

TRADELINE models. • *SUPER TRADELINE model.*

Order Number	Range[b]		Features	Includes
	F	C		
•T87F1859[a]	40 to 90	4 to 32	Replaces heating, cooling, and heat-ing-cooling models (use Q539 for heating-cooling). Alternate series 20 terminal markings on 137421A Wallplate.	6 in. [152.4 mm] cover ring, 137421A 3-terminal Wallplate.
T87F1867	50 to 90	10 to 32	With positive OFF.	6 in. [152.4 mm] cover ring, 137421B 3-terminal Wallplate.
T87F2360	45 to 75	7 to 24	Limited set point range.	104456B Wallplate.
T87F2782	35 to 65	2 to 18	Heating only; limited set point range.	Locking cover and locking dial. Al-len-head screws and wrench included for locking cover. 137421A 3-terminal Wallplate.
T87F2873	40 to 90	4 to 32	Concealed 2-terminal heating only wallplate.	104456B Wallplate.

[a]Use T87F1859 as a replacement for former TRADELINE model T26A1433 and T87C1252.
[b]Temperature scale in Fahrenheit on thermostat.

TRADELINE

Fig. 14-2. Catalog sheet for T87F Thermostat, giving specifications. (Courtesy Honeywell Inc.).

Fig. 14-3. Honeywell Q539-2 subbase for use with Honeywell T87F thermostat. (Courtesy Honeywell Inc.).

arrange them so that the stat can be pulled away from the wires and off the wall (after removing the mounting screws). If the hole in the wall is open, stuff fiberglass insulation in it to prevent drafts from affecting the operation of the stat.

The new stat is round and there are level lines on the backplate that is to be mounted to the wall with the screws furnished. This back-plate must be level because the stat has a small mercury-filled bulb that acts as the contact points instead of open contacts. If the stat is not level, the set point (degree mark, 68° F or 70° F, at which the stat pointer is set) will not reflect the actual operating point because of the stat being off level.

Use a small pocket level to give you the horizontal line. Level the base. While holding the base steady, mark the three mounting screw holes with a pencil. These holes are elongated on a curve to allow level adjustment. While still holding the base in place, recheck with the level. Measure twice; cut once. As an example, an opening cut in either drywall or plywood in the wrong place might cause you to have to use another new sheet of either material. Figure 14-4 shows how to wire the T87-F stat.

If you use a very small drill to drill into the drywall, you do not need screw anchors. Just use the screws furnished. For this type of drilling (small-size drill bits) in plaster and the like, I use the Yankee nickel-plated push drill. This drill seems to always be part of telephone installers' tool kits. It is very handy; mine is 40 years old, and it still works.

Mount the stat base on the wall and tighten the screws. You will have fed the three wires through the hole in the stat base. The red wire can be taped up and pushed back through the wall to get it out of the way. White and blue are the only ones used. Either wire goes on either terminal screw. It makes no difference. On other than two-wire stats, a color code is very, very important. After you have the base wired, mount the stat body on the base with its three captive screws.

The stat heat anticipator must be set according to the rating of the primary control. See Fig. 14-5. This control is usually a gas valve,

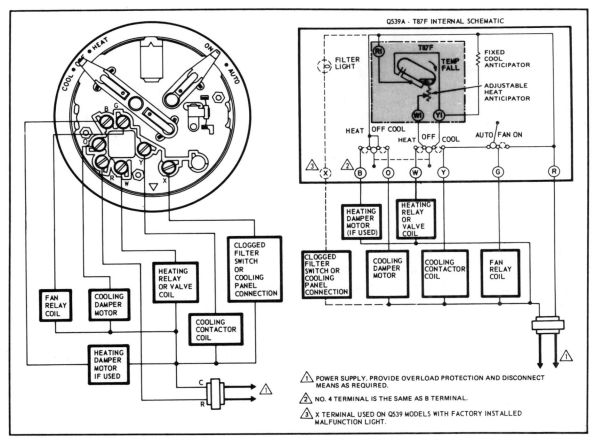

Fig. 14-4. Wiring diagram for use with T87F thermostat and Q539 subbase. For heat or cool and heat/cool applications. (Courtesy Honeywell Inc.).

Fig. 14-5. Close-up of heat anticipator scale and pointer.

Gradually raise the set point until the furnace starts. Now leave the stat alone unless the setting is too high for your needs. Adjust the set point as you prefer. Monitor the new stat. Calibration is seldom needed. If it is needed, detailed instructions are furnished with the stat.

MOVING THE STAT

The installation of a stat in a different location is the same as installing any new electrical device in a hollow wall. Select the new location bearing in mind the most suitable place. Drill a pilot hole down through behind the base shoe molding, or drive a long, thin nail or spike. This will give the location for drilling up from the basement or crawl space. It is also necessary to measure very carefully where to drill down from the attic. Be careful not to step through the ceiling. The other way would be to drill a very small hole at the point where the wall meets the ceiling. Drill upward at an angle similar to locating for the basement drilling. This hole is easily patched with spackling compound.

Stats are mounted five feet above the floor. Drill or pound a ½-inch hole in the drywall. It is easy to use a ⅛-inch wide screwdriver and your electrician's side cutters (pliers). Hammer the screwdriver handle with the pliers. This operation will hurt neither tool because the drywall is very soft.

If you have already drilled up or down to get inside the wall, drop a mouse (solder formed into a weight, tied to a cord) down from the stat hole or down from the attic. Fish out the mouse with the "fish hook" described in Chapter 13.

If the new location is farther from the furnace than the old location, you might need more thermostat cable. This cable consists of three or more No. 18 copper wires in a covering similar to Romex cable. The additional length may

relay, or an oil burner control. The stat wires are connected to this control. The current rating of the control is printed or stamped on the body and is a number such as 0.2, 0.4, or 0.45. The heat anticipator is the scale with the small pointer directly under the mercury tube of the stat. The scale has markings corresponding to the primary control markings. Set the pointer at the control rating. Install the cover and set the stat at 68 degrees. Go to the furnace and turn on the furnace switch. If the furnace does not start, the stat set point was lower than the room temperature and the furnace will not start.

be spliced. A junction box is not required for low-voltage splicing. You can also run all new cable if you prefer. If the cable is spliced, use wire nuts. Always splice color to color if the individual wires have colored insulation. Very old cable might have all-black wires. Be sure to staple the cable so that it will not be damaged. Keep it close to the framing members. Sometimes you want to drill holes to run it through framing members. Keep it neat. Insulated staples are available at hardware stores.

NIGHT SETBACK WITH TWO STATS AND A TIMER

Using two stats and a time clock (timer) saves you money over the standard night setback stat sold in hardware stores and home centers. Stats cost about $15 to $20 and a timer costs about $25. The store price of the fancy model is usually $80 to $90. You also save on installation labor. See Fig. 14-6.

Your original stat can be used as the night stat and put in a hallway. The new stat is in-

Fig. 14-6. Timer to operate night setback, using two thermostats. (Courtesy Intermatic Inc.).

stalled in place of your old one. Stat cable from each stat is brought to a common spot in the basement where the timer will be located. A hallway location for the night stat is desirable because two stats side by side in the living room would look strange. They may be installed side by side; wires from both still must go to the timer.

You need to modify the timer switching mechanism to isolate the *low* voltage (24V, which is to operate the clock of the timer only) from the line voltage. One wire between the two stats is broken by the timer contacts. The two stats are wired in parallel. The wire that is broken (disconnected) actually disconnects the day stat so that it cannot control the furnace. This action usually takes place at night, around 11:00 p.m.

The night stat is usually set at 6 degrees to 8 degrees below the set point of the day stat that is set at perhaps 68° F. This higher day set point keeps the night stat satisfied so that it never calls for heat during the day. At a set time, the night stat is in control and does its job. It keeps the house at 60° F or whatever its set point is. At perhaps 6:00 a.m., the day stat is reconnected in the circuit and then maintains the daytime 68° F temperature.

The timer you have to use may or may not have to be modified. Intermatic Incorporated manufactures a timer that has the timer motor terminals isolated from the switching terminals. The timer is their Model T-101B(a). This is ideal, because no modification is necessary. This arrangement seems quite difficult, but in reality it is very simple. Usually a home improvement store can order the Intermatic timer for you, or an electrical wholesaler might sell you one.

One disadvantage with this type of system is that there is no visible clock on either stat.

This might or might not be something that is desirable to you. Generally, timepieces are selected for their appearance and there is really no need for another one. The timer has to be wired to a source of power that cannot be interrupted at any time. This is a standard type of installation and is no different from the wiring of a receptacle. Just make sure that the power supply is uninterrupted.

Should a utility power interruption occur, you must reset the dial on the timer. Open the cover and turn the dial manually to the time of day. Check all timers for correct time of day occasionally as power outages at night may not show up unless you have an electric clock that has no battery carryover.

The day-night thermostat described in the following section does have the battery carryover feature. Other night-setback stats are on the market. Some have expensive, exotic features that are not needed. Be a prudent buyer and buy the best you can afford (but try to buy when on sale).

THE STANDARD NIGHT-SETBACK STAT

Many manufacturers of controls offer night setback stats. See Fig. 14-7. I recommend the Honeywell T-8600-R-3 Fuel Saver Thermostat. See Fig. 14-8. This stat comes complete with instructions. Wire all terminals correctly, color to color, or according to instructions.

The T-8600-R-3 stat has a carryover feature so that the clock continues to run in the event of a power outage. The heat anticipator scale pointer must be set as described in the previous section to match the amp rating of the primary control that it will operate. The instruction booklet provided is complete and it shows wiring diagrams for almost any stat to be replaced.

Fig. 14-7. Honeywell T-8082A thermostat with cover removed. Shows night set-back dial and control mechanism.

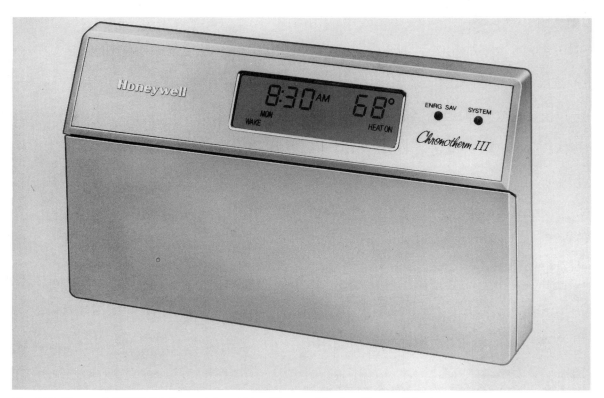

Fig. 14-8. Honeywell T8600-R-3 thermostat complete with cover. (Courtesy Honeywell Inc.).

I recommend that any device purchased be thoroughly inspected before leaving the store. A defective device means a trip back to the store. Save money by saving needless trips. Sometimes parts are broken or missing. If you find the stat OK, remove the shipping tabs. The stat will not operate with these in place.

Take your time, work carefully and accurately, and check your finished work. I once checked out from the company stock a new brass valve for a gas line shut-off. After screwing it on the gas line, I tried to screw a nipple into the opposite end. That end had never been threaded at the factory. This forced me to drive to a wholesale supply house to obtain a new valve. You can be sure I inspected both ends of this second valve. This not only caused loss of time, but it made necessary the relighting of numerous pilot burners on other gas unit heaters.

PILOT LIGHT TO INDICATE OPEN GARAGE DOOR

Garage doors that are operated by remote control are sometimes left open. An indicating pilot light will alert you to this condition. An open garage door after dark is an invitation to vandals. See Figs. 14-9 through 14-13.

Garage door operators usually light an integral light when the door is opened. This al-

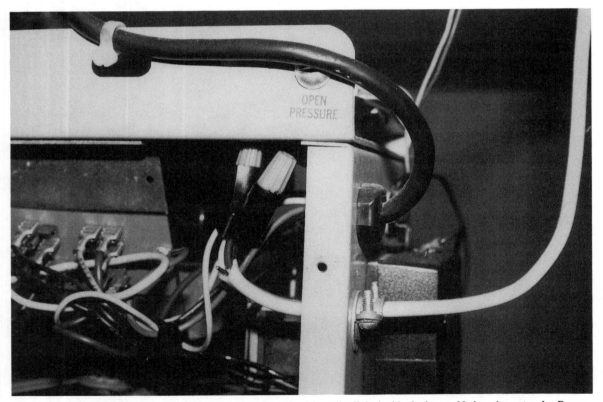

Fig. 14-9. Connection made to replace light on garage operator with a pilot light inside the house. Notice wire nuts, also Romex leaving housing of operator.

Fig. 14-10. Pilot lights mounted on the ceiling. Left, garage ceiling light. Right, indicates "OPEN" garage door when lighted.

lows you to get out of the car and remove packages in a lighted garage. Other door operators, after a short time, turn off the light. A pilot light can be installed with either the continuous light or the delayed turn-off light.

To proceed with the installation, determine where you want to locate the pilot light. I have mine installed in the ceiling over the door lead-

ing out to the garage. A convenient wall is also suitable. The logical place is where the light will be easily noticed. For the installation, you need enough Romex to reach from the operator power unit to the pilot light location. Run this cable as you would any electrical circuit.

The cable should be run up to a ceiling joist or above the finished ceiling over to the light location. Leave enough at each end for proper connections. Also needed is a wall box and a cover. The cover plate depends on the type of pilot you use. There are many types of pilot lights on the market:

▶ The oldest type has a small (7½W) bulb and socket. These take up a whole wall box alone.

▶ The interchangeable line (Despard brand) uses a 7½W, the same bulb that extends

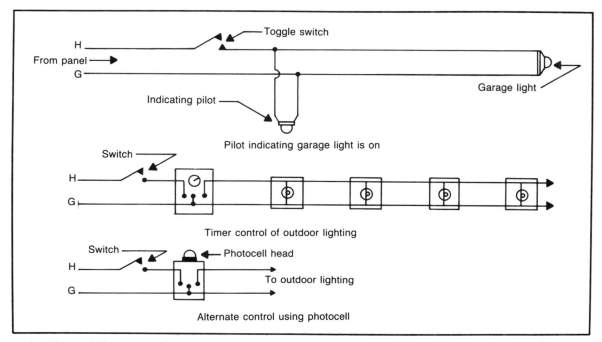

Fig. 14-11. Installation of pilot light to indicate the garage light is on. Timer control of outdoor lighting. Photocell control of outdoor lighting.

Special Projects

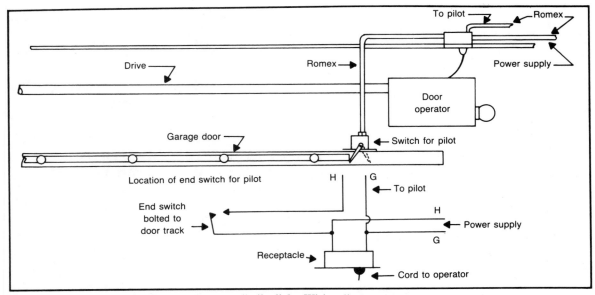

Fig. 14-12. *Location of end switch for "door open" pilot light. Wiring diagram for track mounted switch.*

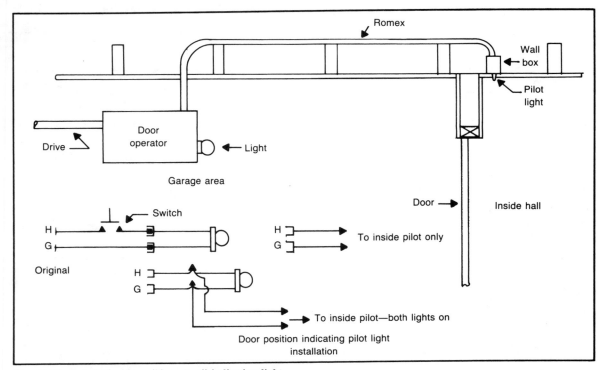

Fig. 14-13. *Installation for a "door open" indicating light.*

through the cover plate with a metal guard over it. A newer model of the same style uses a neon bulb and is flush with the cover plate.

▸ A third type is sold by Radio Shack and is used in electronic equipment. It is rated at 120V. I used this type for my lights. A blank plastic cover plate has to be drilled for this light. The light is pushed through a snug fitting hole in the plate and held by a metal friction ring on the back. The two leads are connected to the Romex wires in the box.

▸ Another type uses a duplex receptacle and plate and a plug-in night light socket and bulb. One on the market is very small. It is about the size of a small attachment plug. The neon bulb is inside a translucent plastic cover. This type is very convenient because the light is easy to change. Unplug and insert a new one.

Unplug the door operator from the ceiling receptacle, then remove the cover from the power unit. Trace the leads (wires) from the light socket in the power unit to their connections at terminal screws or push-on connectors. If you want this light to continue to operate as before, you will have to splice into the leads between the socket and the terminals. Cut one wire at a time and splice in the Romex wires, white to white, black to black. Each color will have three wire ends to splice together with a wire nut.

Install the wall box and run the Romex to the operator power unit. There might be an unused knockout in the cabinet that you can use. If not, you will have to make one. Use a hole saw that is ⅞ inch in diameter. Be sure there is nothing inside where you will be drilling. Use a Romex clamp in the knockout. Connect the Romex wires to the light leads or directly to the

terminals if you don't use the light. Mount the pilot light wall box where you prefer and run the Romex into the box. Connect the pilot light you have selected and install the cover plate. Recheck all of your work carefully and make sure nothing is left undone.

Plug the door operator cord back in the receptacle. If the door is open, the light should come on. If not, recheck the wiring connections. If nothing operates, check the circuit. Use a bulb adapter to test the receptacle. You should now know how to test fuses and circuit breakers. There should be no trouble finding the problem.

If the door operator turns its light off after a short period, you can bypass this delay mechanism. If you cannot do this easily, it is necessary to devise another means of energizing the pilot light. A limit switch can be installed on the door track at the far inside end where the top of the door stops. Mount the limit switch so that its lever is moved and closes its switch. Instructions with the switch explain how to make this adjustment. You can get power from the receptacle where the door operator is plugged in. The expensive item is the limit switch; the list price is about $18. You might be able to get a discount at the wholesaler or perhaps find one at a surplus store. Make sure the surplus store unit operates properly. The limit switch must have the terminals enclosed; the NEC requires this. Also check to see if the actuating arm makes contact when pushed and returns by spring action when released.

Wiring consists of a two-wire cable from the limit switch to the power source (junction box), and then to the pilot light. While most operators are plugged into a receptacle, some might be wired directly to the circuit. This is no problem when there is a junction box near the operator. You will find an unswitched circuit there from which to get power.

Fig. 14-14. Toggle switch to disconnect radio receiver connection to door operator when on vacation or lengthy absences. When this switch is off no stray signal (from another car) can open the door. A key-operated switch is needed near the door, outside, so as to operate the door when this switch is off or the car is away.

In addition to controlling the door operator with the radio sender kept in the car, there might be times when you would like to close or open the door from outside the garage. When all family members are away, a neighbor might need to have access to the garage. In such a case, it is necessary to provide a key-operated switch on the outside of the garage, usually on the door frame. See Figs. 14-14 and 14-15. Door operator manufacturers furnish these key switches. To install a key switch, locate the wires from the inside push button (this closes or opens the door from inside the garage and out of the car). These two wires will be in the garage attic or above the door track area, either singly or twisted together. Usually, a ¾-inch hole needs to be drilled for the key switch. Because the voltage is low, the wires can be concealed in slots on the inside of the door jamb and led up to where the push-button wires are located. Pull the plug of the operator before working on the control wiring. This is to prevent the door from going up and down while you are working on the wires.

Drill the hole for the switch. Run new wires from the area of the push-button wires down to the key switch hole you have drilled. You can use bell wire or thermostat cable. Cable is better because it has a heavy covering over both wires for protection. When concealing the cable (or wires), be sure they are out of the way of any construction work done later in the garage. You might drive a nail and have the garage door open.

Connect the wires to the two terminals of the switch. Push the switch back into the hole you drilled as you pull on the wires to take up the slack—no slack is needed here. Continue taking up the slack and keep the wires snug and

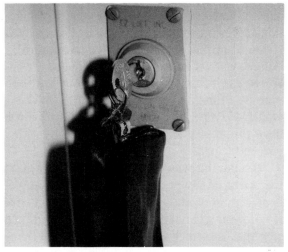

Fig. 14-15. Key operated switch to operate door when car with transmitter is away, or receiver is shut off by switch shown in Fig. 14-14.

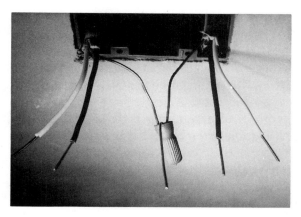

Fig. 14-16. *Wires for two different uses prepared for connection. The left two wires are for a toggle switch to control a ceiling fixture. The right two wires feed a duplex receptacle. The bare wires are to ground the receptacle and possibly the switch (not all switches have a grounding screw at this time).*

hidden in their groove or whatever route you take. Staple where necessary.

When you have the new wires adjacent to the push-button wires, cut one push-button wire if you have slack enough to provide for a pigtail splice. Bare both cut ends, plus one end from the new key switch pair and make a pigtail splice of all three (making it tight with a wire nut). If you have no slack, then bare a 1½-inch section and make a tap splice. Solder it if you like (that's the best way) and tape the joint. Don't forget to do the other wire the same way; it won't work with one wire.

PILOT LIGHT ADDED TO EXISTING SWITCH

A room or area that has the light switch remote from that room, such as outside a closed door, should have a pilot light to indicate when that light is on. Likely areas would be the basement or attic. Both these areas often have the switch located in the hallway outside near a closed door to that area. Two arrangements are possible. If the present switch is visible and is outside the area, it can be combined with the pilot in the existing single-gang box. This is a

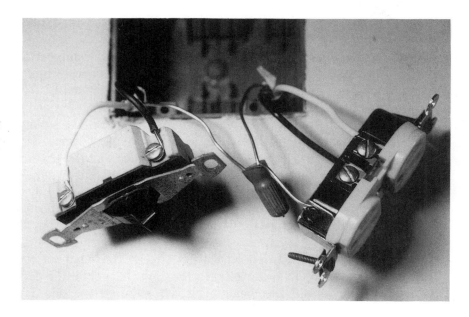

Fig. 14-17. *The switch and receptacle have been connected. The ground is connected to the receptacle.*

Fig. 14-18. The switch and receptacle have been slanted slightly as the box has been installed off level.

combination device and costs about $5. The price is reasonable considering that you will gain a toggle switch. This combination uses the standard duplex receptacle cover. See Figs. 14-16 through 14-18.

The second arrangement is where the switch is not visible or is inside the area. This condition calls for another wall box to be installed outside the area in a visible location. It may be necessary to run a length of two-wire grounded cable from the light fixture in the room out to the pilot light location. Sometimes you can make the connection at the switch location. Basement and attic lights are frequently left on all night.

Another area where lights are left on is the front and rear outdoor entrance lights. This is easy to do because the switch is at the door and is ready-made for this change (both the "feed" and the "load" to the outdoor light). There can be a second switch for an inside hall light. These two switches might be in a single-gang box or a two-gang box. A single-gang box poses the problem of assembling a three-device interchangeable assembly. While this crowds the box some, it should not exceed the limit of wires. Two switches in a two-gang box is the perfect condition as the inside switch is left and the outdoor switch is exchanged for the switch pilot light combination. Use a switch-duplex receptacle plate.

CONTROL OF OUTDOOR LIGHTING

Control of outdoor lighting can be accomplished by turning it off by hand, a clock timer switch, or by a photocell. In each case, the wiring is the same, but the operation of each control differs. The photocell turns the lights on at dusk and off at dawn. The timer turns the lights on and off at the times you select. Because dusk and dawn vary during the year, it is necessary to move the settings to correspond to these changes. One special time control called the "astronomical" timer does this for you, but it costs $45 and up. The standard timer is $18. The photocell is about $15. So the astronomical timer is at a serious disadvantage—the "astronomical" price! Timers have a more positive action, but the photocell is entirely satisfactory for the homeowner. Just buy the best you can afford, because many inexpensive photocells are erratic in operation and flash on and off, disturbing neighbors if not the owner. Refer to Figs. 14-19 through 14-25.

You might not now have any outdoor lighting except for the decorative lights at the front entrance and side or rear doors. This lighting

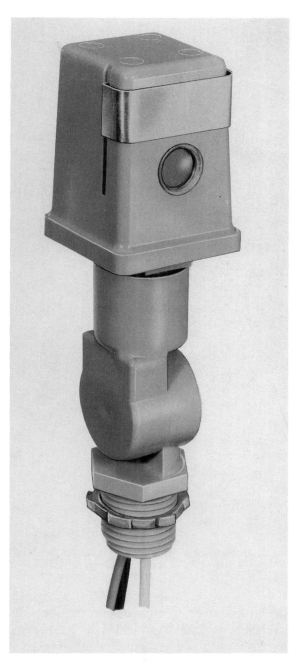

Fig. 14-19. Adjustable photocell used to control perimeter security lighting, Model K-111. (Courtesy Intermatic Inc.).

Fig. 14-20. Adjustable photocell, K4221. Also has an adjustment for light level hitting cell eye as shown. (Courtesy Intermatic Inc.).

Fig. 14-21. Built-in timer for such uses as: attic fan, sun lamp, garage or basement lights. (Courtesy Intermatic, Inc.).

shine along the four outside walls can be either fed from the attic at each corner or light-colored non-metallic cable (Romex) can be run under the eaves to the various fixtures. Use type NMC cable (waterproof) and keep the fixtures under the eaves. Use the special waterproof boxes and covers to mount the outdoor sockets. You can also use Thinwall tubing (EMT), but that is expensive and not necessary.

The installation of the photocell or timer is very easy. In effect, both devices are switches, as is a stat. The timer should be mounted inside to prevent vandalism or damage. There are timers in weatherproof cases, but they are not necessary and are more expensive. If your outdoor lighting was installed some time ago, it has a manual switch to turn it on and off. The timer must go between this switch and the outdoor lighting. This also applies to the photocell location. This is the sequence: power supply, switch, timer (or photocell), lighting.

By its nature, the photocell must be installed outdoors. The north side is preferable but not necessary. As it senses light, it must not be facing any lights such as other controlled outdoor lighting, traffic signals or street lights. Any light ''seen'' by the cell causes erratic action.

ELECTRIC WATER HEATER TIMER

Installing this timer will save you money. Water heaters can be cycled so that water necessary for household use is heated, but the rest of the time is not kept hot. As much as a 33 percent saving on energy costs can be realized. Refer to Figs. 14-26 through 14-35.

Some utilities have the water heater on its own meter and the rate is different from the lighting rate. Power to the heater can be inter-

can be left as is or expanded to include side floodlights or other all-around lighting for security reasons. If you plan on installing extensive security lighting or just mainly decorative lights, decide on a layout and locate the positions for the fixtures. Spot lights on the front corners that

Fig. 14-22. Portable timer with cord and plug. May be used to control any portable equipment within its capacity. (Courtesy Intermatic Inc.).

rupted by a radio signal from the utility. A special device is next to the heater electric meter. If your meter has such a device, you could question the utility about times when power is cut off by radio. Other utilities have their own timers (outside).

If you have made the decision to install the timer on your water heater, first buy the timer. They cost about $30. You also need two more greenfield connectors because you must cut the present greenfield to allow it to go into and out of the timer case. From there, it goes to the heater junction box. If you can visualize the lay-

out, you will be able to tell if the connectors should be straight or angled. You could get two of each. The timer must have a 40A rating; most do if they are sold for heaters. Locate a spot near the heater so that the greenfield will attach to the case easily.

Turn off the power to the heater. Next, test to see if the power really is off. Lock the breaker or remove the fuses and take them with you. Be safe. To disconnect the wires from the heater, remove the cover plate. This is either on the actual top or near the top on the rear side. The wires from the distribution panel end

197

Fig. 14-23. Eight-day on-off timer for commercial uses where Saturdays, Sundays, and holidays may be skipped and allowed to remain on their night settings.

at screw or nut terminals in the small area under the access plate. Some heaters have an external octagon box as a terminal cover. Other heaters just have wire nuts connecting the heater wires to the wires from the fuse panel.

Be careful when loosening the terminals because they might be corroded and you might break some plastic part. Measure and cut the greenfield. Make sure you saw at right angles to the spiral and not straight across the length. You should attach the box connector before cutting for a more accurate measurement. If you

have enough length left, you can use this piece to go from the timer case to the heater. Sometimes the wires in the greenfield are long enough; other times they are too short. It is better to buy 10 or 15 feet of No. 10 copper and have enough slack to make proper connections. Allow for neat loops rather than tight, short connections.

Basement heaters have the wiring coming from the ceiling. You should be able to install the timer by the distribution panel and connect it by means of an offset nipple. You will not have to touch the heater connections. First-floor heaters usually have the greenfield coming out of the wall. In this case see the preceding paragraph. If you need extra wire, use black for the hot and grounded wires.

Paint the grounded wire white at both ends or use white tape. Strip the insulation for the equipment ground wire. Use white wire if you can buy it. When you install the box connectors, be sure to tighten the locknuts securely to complete the equipment ground continuity of the metal housing and greenfield.

Remember, if you have a 240V water heater, you do not have a white wire; you have two black wires and one bare wire for grounding the noncurrent carrying parts such as the timer case, greenfield, etc. A 120V heater has one black wire, one white wire, and one bare wire.

When you have rerouted the greenfield, you can now pull the wires in. In the timer, the terminals will be marked "line 1, load 1; line

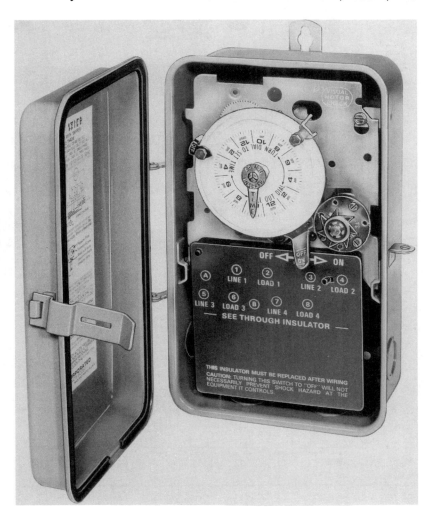

Fig. 14-24. Domestic model of the eight-day timer in Fig. 14-38. Will skip any days selected. (Courtesy Intermatic Inc.).

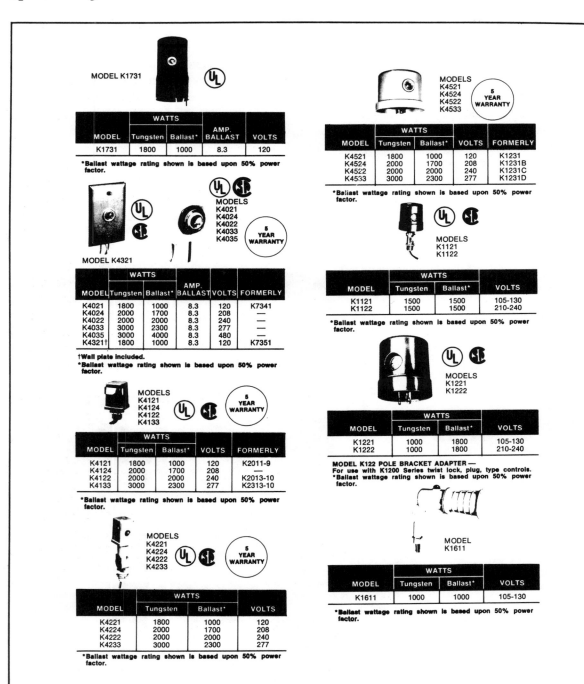

MODEL K1731

MODEL	WATTS		AMP. BALLAST	VOLTS
	Tungsten	Ballast*		
K1731	1800	1000	8.3	120

*Ballast wattage rating shown is based upon 50% power factor.

MODELS
K4021
K4024
K4022
K4033
K4035

MODEL K4321

MODEL	WATTS		AMP. BALLAST	VOLTS	FORMERLY
	Tungsten	Ballast*			
K4021	1800	1000	8.3	120	K7341
K4024	2000	1700	8.3	208	—
K4022	2000	2000	8.3	240	—
K4033	3000	2300	8.3	277	—
K4035	3000	4000	8.3	480	—
K4321†	1800	1000	8.3	120	K7351

†Wall plate included.
*Ballast wattage rating shown is based upon 50% power factor.

MODELS
K4121
K4124
K4122
K4133

MODEL	WATTS		VOLTS	FORMERLY
	Tungsten	Ballast*		
K4121	1800	1000	120	K2011-9
K4124	2000	1700	208	—
K4122	2000	2000	240	K2013-10
K4133	3000	2300	277	K2313-10

*Ballast wattage rating shown is based upon 50% power factor.

MODELS
K4221
K4224
K4222
K4233

MODEL	WATTS		VOLTS
	Tungsten	Ballast*	
K4221	1800	1000	120
K4224	2000	1700	208
K4222	2000	2000	240
K4233	3000	2300	277

*Ballast wattage rating shown is based upon 50% power factor.

MODELS
K4521
K4524
K4522
K4533

5 YEAR WARRANTY

MODEL	WATTS		VOLTS	FORMERLY
	Tungsten	Ballast*		
K4521	1800	1000	120	K1231
K4524	2000	1700	208	K1231B
K4522	2000	2000	240	K1231C
K4533	3000	2300	277	K1231D

*Ballast wattage rating shown is based upon 50% power factor.

MODELS
K1121
K1122

MODEL	WATTS		VOLTS
	Tungsten	Ballast*	
K1121	1500	1500	105-130
K1122	1500	1500	210-240

*Ballast wattage rating shown is based upon 50% power factor.

MODELS
K1221
K1222

MODEL	WATTS		VOLTS
	Tungsten	Ballast*	
K1221	1000	1800	105-130
K1222	1000	1800	210-240

MODEL K122 POLE BRACKET ADAPTER —
For use with K1200 Series twist lock, plug, type controls.
*Ballast wattage rating shown is based upon 50% power factor.

MODEL
K1611

MODEL	WATTS		VOLTS
	Tungsten	Ballast*	
K1611	1000	1000	105-130

*Ballast wattage rating shown is based upon 50% power factor.

Fig. 14-25. Selection of photo controls. (Courtesy Intermatic Inc.)

Fig. 14-26. Timer for an electric water heater. Can save up to 33% on electric usage. Lever in cover can be used to turn heater on for additional hot water. (Courtesy Intermatic Inc.).

The ON/OFF lever may be in the OFF position. While this lever is OFF, set the time of day and the ON and OFF trippers to the times you want the heater to be on and off. Usually two ON periods, morning and evening, are sufficient for the average family. Figures 14-26 and 14-27 show a representative timer. The mechanism is removable for installation purposes or for replacement. Now you will be saving money on your electric water heater operating expense. One additional money saver tip is to install a water heater insulating blanket on the outside of your heater. You can feel the difference just by touching.

Note: If the utility can interrupt the power to the water heater, you need to run a separate independent circuit to operate the clock motor of the timer. This is necessary to keep the timer "on time" to correspond with the time of day. It is a good safety practice to put a warning notice inside the timer case (over the fiber insulator), noting that there are two separate power supplies to this device. Turn off both circuits before working on the timer.

WIRING A DETACHED GARAGE OR OUTBUILDING

Even though many garages are now attached to the house (a great convenience), there are thousands set back in the rear of lots, unlighted, separate from the house. Save money; wire you own garage. An outbuilding should be wired underground. See Fig. 14-36. This eliminates low overhead wires, with their dangers and maintenance. Use underground cable type UF (underground feeder). See also Figs. 14-37 through 14-39.

I recommend 3 No. 10 UF cable to your garage or outbuilding. Try home improvement centers and hardware stores for the cable. Near

2, load 2." Line is from the power supply and load is to the current using device (in this case the heater). Do not try to skimp on either the greenfield or the wires. Buy more rather than try to make do. Route the wires neatly in the timer case with a nice "S" loop.

If everything is ok, restore the power (put the fuses back or turn on the breaker). Check for power at the heater and timer. There is an ON/OFF lever for manually operating the timer. Do not use this lever to disconnect power from the circuit. Use the fuses or breaker to disconnect power, *always!*

ELECTRIC WATER HEATER TIME SWITCH
SERIES T104-20

- Can be set to shut electric water heater off during periods of your utility's peak power usage.
- Time switch can be automatically programmed to meet the needs of individual family.
- Save up to 33% of your water heating dollars by reducing hot water temperature and quantity.

Electric water heaters consume more energy than any other appliance. The average homeowner spends $216.00 per year on electric water heating costs (based on 4500 kilowatts per year, a utility rate of 4.8¢ per kilowatt hour). With energy costs on the rise, there is now a way to help you reduce water heating costs. The water heater, unless timer controlled, automatically maintains the hot water temperature set by the homeowner, night and day, whether needed or not. By automatically limiting water heater energy usage to those hours needed, the homeowner can reduce energy consumption and save money.

For maximum savings, set timer to turn on one hour in the morning and two hours in the evening. Set trippers to turn water heater on one hour ahead of expected morning and evening periods of major hot water usage. This will normally provide sufficient hot water for the average family and will cut the electric water heating bill by an average of 33%.

SPECIFICATIONS

CASE — Drawn steel 7¾" (19.7cm) high, 5" (12.7cm) wide, 3" (7.6cm) deep in gray finish. Spring hasp, with hole for lock, holds permanently attached side hinged door closed. Three mounting holes on back.

KNOCKOUTS — Combination ½" — ¾" nominal knockouts, one on back and each side of case, and two on bottom.

SPECIAL VOLTAGES AND CYCLES — All models 125v, 50 Hz; 250v, 50 Hz.

SWITCH RATING — 40 max. load amps. per pole, 35 amps. tungsten; 2 H.P., 125V. A.C. and 3 H.P., 250V. A.C.

MODEL NUMBER	SWITCH	VOLTS 60 Hz	AMPS. PER POLE INDUCTIVE
T103-20	DPST	125	40
T104-20	DPST	208-250*	40

For additional trippers, order No. 156T1978A.

Fig. 14-27. Water heater timer specifications and mechanism. (Courtesy Intermatic Inc.).

where I live there is a pipe and supply company (actually a large hardware store) that has everything the homeowner could want in electrical supplies. In such a place you will be able to buy the length you need. Be sure to buy enough. Splices are prohibited underground.

The No. 10 three-wire cable allows you to have 240V power in the building which allows you to operate power tools with little voltage loss. This cable must have overcurrent protection (fuses or breakers) at the starting point, such as at the distribution panel in the basement or other location. Use 30A overcurrent protection in the house panel. If you use 30A protection there, you need to install a fuse/breaker panel in the outbuilding to protect the smaller wires used there.

If you do not want to have a fuse/breaker

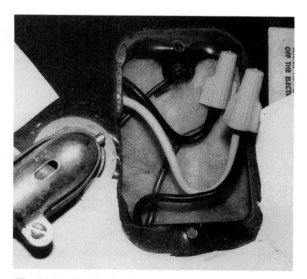

Fig. 14-28. Junction box for connection of supply wires from timer. There are no terminals, connections are made using wire nuts. CAUTION: Be sure power is off before starting work.

If you use No. 10 UF cable underground and provided protection for 30A, you must also use No. 10 Romex or BX between the distribution panel in the basement and the junction box at the outside wall where the UF cable starts. If you protect the UF cable with only 20A, then use No. 12 Romex or BX in the basement. Never overfuse any part of a complete circuit. As the smallest wire is the weakest link, the overcurrent protection must protect the weakest part. The weakest link in any circuit must be the fuse or breaker. This is the fail-safe link; its duty is to fail when necessary.

The NEC requires that where underground cable rises out of the ground, it must be protected with conduit or Thinwall (EMT). The method used is to run conduit through the house

Fig. 14-29. Heater thermostat dial (under cover plate). Turn off power first. Make adjustments using a small screwdriver.

panel in the building and are using No. 12 or 14 Romex, protect this circuit at the house (Fig. 14-38.) It must be protected by 20A or 15A overcurrent devices respectively. You can also use No. 12 three-wire UF cable and fuse it in the house panel at 20A, then the whole length of the circuit will be protected by one fuse/breaker.

Note: If the distance from the house to the garage is great, it is advisable to use No. 10 UF cable. This will avoid any voltage drop in this long line to the garage.

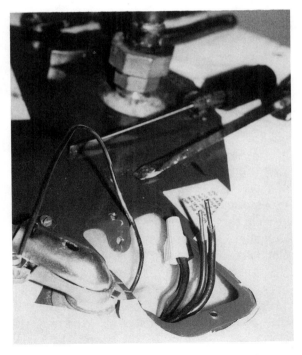

Fig. 14-30. Power leads have been disconnected from heater to start installation of timer.

arranged to go down into the ground at least 12 inches. This piece can be straight or have a 90-degree bend pointing toward the outbuilding. A bushing must be on the end that is in the ground. The trench must be at least 12 inches deep. Snake the cable in the trench. Do not pull it taut. This might look sloppy, but it prevents tension on the cable that could damage it.

All wiring must be neat and not be stretched out taut. Some, when wiring receptacles or switches, leave no slack, even for pulling the

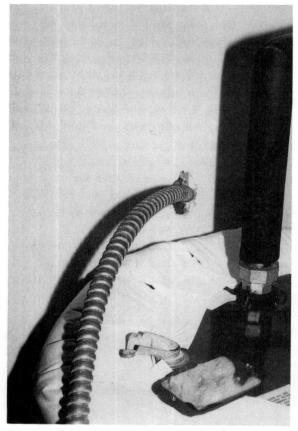

Fig. 14-31. Greenfield has been disconnected from heater connector.

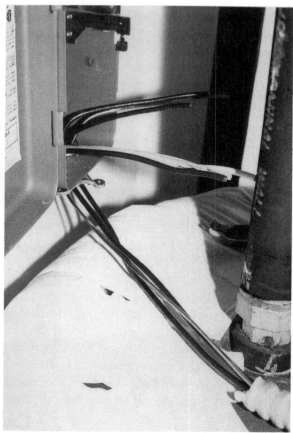

Fig. 14-32. After pulling greenfield from wall, wires are found to be No. 10 Romex. Timer mechanism has been removed for easy wiring. Black tape will be put on white wire, 240 V circuit does not require a white, grounded wire.

wall from a junction box mounted on the side of a joist (this is to make the change from other types of cable to the underground UF cable).

On the outside is a special fitting called a Condulet. The position of the openings in this Condulet is designated by letters, in this case LB, meaning L=left; B=back (angle fitting, back outlet, and bottom outlet). (See Fig. 14-36.) This LB is connected to the conduit with the compression connector (part of the Condulet). See Fig. 14-36.

Another piece of conduit is attached to the downward-pointing opening in the Condulet and

Fig. 14-33. *Hacksaw cut in greenfield. Twist and flex to separate.*

device out of the box (they must have wired the device, then gone around to the back of the box and pulled the device into the box by pulling on the Romex cable). I have seen devices that looked exactly like this. It would take a feat of magic to disconnect such a device!

Wiring the garage is just like wiring any building that does not have the inside finish on the studs (some garages do have drywall on the studs). You want two or more ceiling lights, depending on the positioning of vehicles, and perhaps a workbench to the side or rear. If you intend to build a shop in the garage or other building, do run No. 10 copper cable all the way from the distribution panel in the basement to a fuse panel in the garage. Use No. 10 three-wire interior Romex and No. 10 three-wire UF cable in the ground.

Two or more receptacles over the workbench should be installed. A neat way to do this

is to use handy boxes with short (2-foot) lengths of ¾-inch Thinwall with appropriate box connectors separating the boxes. This gives you six receptacles over the workbench.

You can even have two separate circuits in this arrangement. To do this, install four boxes and three conduit lengths. Put alternate receptacles on different circuits. Separating the individual outlets on the duplex receptacle is unnecessary and I recommend boxes 1 and 3 on one circuit and boxes 2 and 4 be on another circuit. If you have a 240V appliance or welder, provide a special 240V receptacle for this. An

Fig. 14-34. *Wiring has been completed to timer. Note that the white wire on the far right terminal is identified as black using black tape. The grounding wires are connected to the box screw in the center rear. Replace the fiber safety cover over the terminals when finished.*

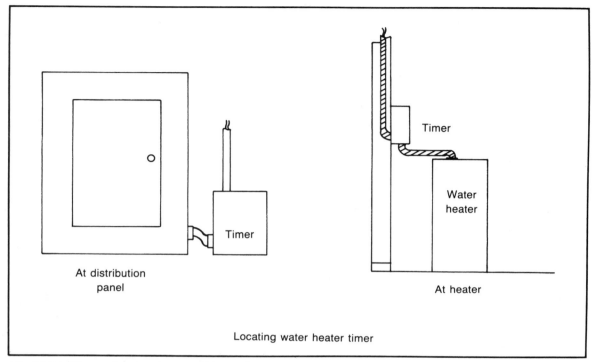

Locating water heater timer

Fig. 14-35. Alternate locations for a water heater timer.

electric dryer 240V receptacle will be ideal for this application. Be sure to buy a receptacle with the correct configuration such that no attachment plug for an appliance intended for 120V only can be plugged into this receptacle.

Fuse panels are sold with the fuseholders already installed. Breaker panels are bare (service-entrance panels usually have the main breaker). Buy either type panel so that you have space for four circuits: two for the 240V circuit and the other two for two 120V circuits.

I had been looking for a two-position breaker cabinet. Prices ran from $12 to $14. I did not buy because I was in no hurry. Later, in a local retail hardware store, I found exactly what I wanted—Sylvania brand at $8.39 (brand new).

Be sure what you buy has the UL label and is new or in perfect condition. Find out if you can return the item if necessary. Circuit breakers should always be purchased brand new in order to eliminate trouble. It would be better if for the 240V circuit the breaker would be "two-pole" with both handles bar-tied together (both poles of the 240V circuit trip at once). This type will take two spaces in the cabinet. One-pole breakers come in ½-inch thick and 1-inch thick sizes. My preference is the 1-inch breaker. It even looks stronger!

Garage builders sometimes provide a piece of curved ¾-inch conduit coming up through the concrete floor, through the plate, and up into the stud space of the wall. This allows underground wiring to the garage without the need

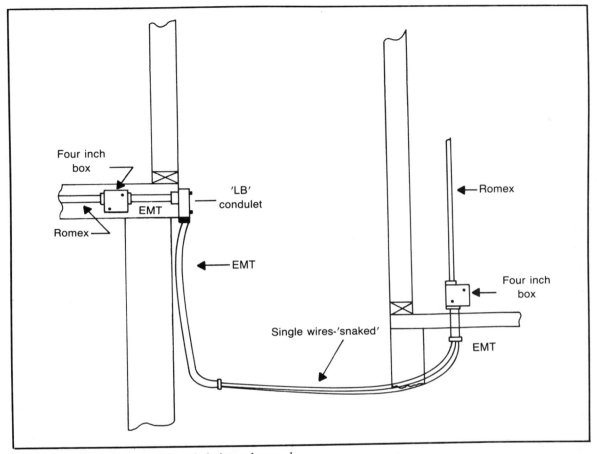

Fig. 14-36. Wiring a garage by running feed wires underground.

for an arrangement such as at the house end. Install a bushing at the below-grade end of the Thinwall and mount a 4-inch junction box on the end inside using a box connector. I had this arrangement in a garage of mine. From this box, I ran Thinwall up to a 4-circuit fuse panel.

AUXILIARY POWER SUPPLY

The use of auxiliary power has become more prevalent as certain electrically operated equipment necessary for life support and back- up facilities is being used. This equipment is not only used in hospitals and nursing homes, but also in private homes. For such equipment to be dependable, it must have a constant power supply. Such a power supply can be provided by an auxiliary power source.

Auxiliary power (sometimes called standby power) is generally available in the form of a portable generator. Such a system, consisting of the generator, a transfer switch, and the related wiring can be installed, by yourself, for about $1500 to $2000. Units range in capacity

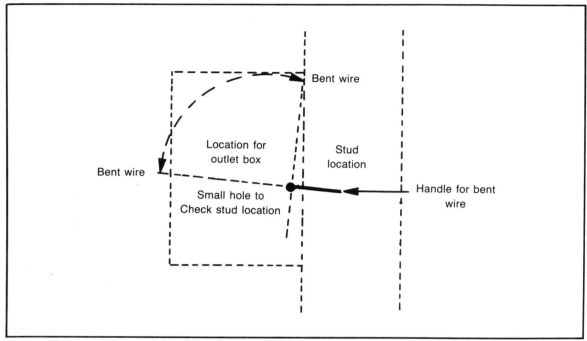

Fig. 14-37. Diagram of method of locating stud so that a large hole is not made in drywall and a stud is encountered there. A bent wire is rotated through a hole to locate the stud.

from 1350W to 5000W for this price. Because the setup is basic, greater wattage does not increase the cost in proportion, therefore, you might want to buy a greater-capacity system. Wiring is simple. The running of the conduit or Thinwall is the hardest part of the job. The standby connection box and cord are on the outside near the meter (as is the standby generator location). See Figs. 14-42 through 14-44.

The wires and conduit from the connection box lead inside to near the service-entrance equipment (as do the wires and conduit from the utility meter). Both supplies, the standby power and the utility power, terminate at a *transfer switch* in a cabinet. This transfer switch can connect either the standby or the utility power to the service-entrance equipment, but not both at the same time.

Fig. 14-38. Cut-away of a circuit breaker. The breaker operates on temperature, amperage, and length and amount of overload.

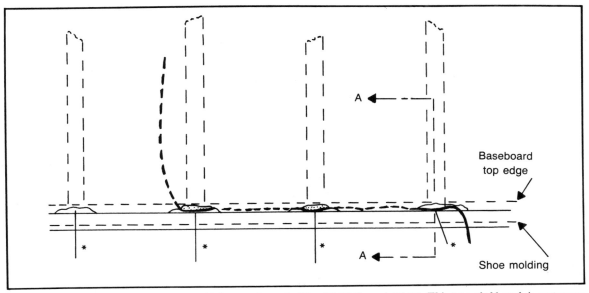

Fig. 14-39. *Routing of cable for the two wall fixtures shown in Figs. 14-28 through 14-35. This type of old work is customary in finished buildings, new or old.*

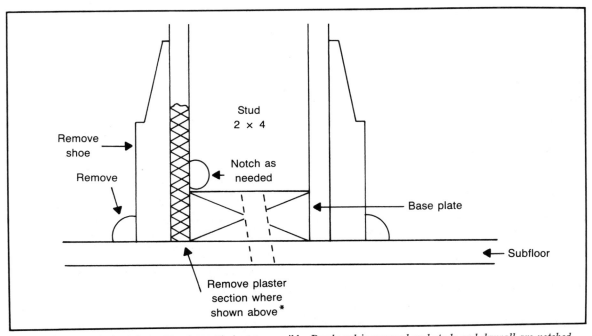

Fig. 14-40. *Method of routing cable when attic is not accessible. Baseboard is removed and studs and drywall are notched.*

209

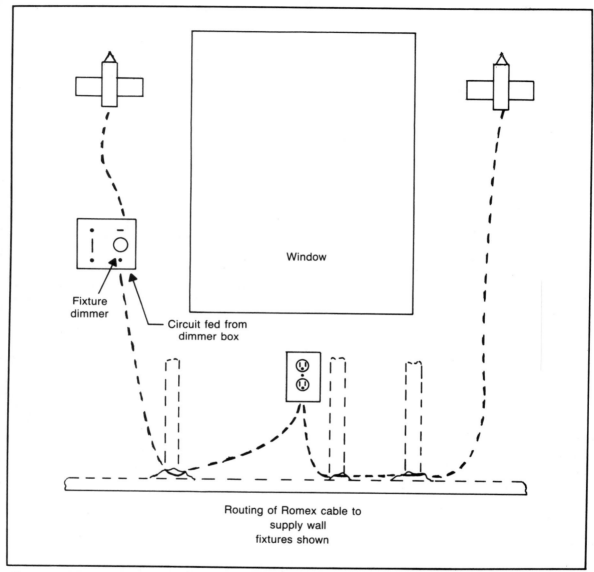

Fixture
dimmer

Circuit fed from
dimmer box

Window

Routing of Romex cable to
supply wall
fixtures shown

Fig. 14-41. Routing of cable to supply power to wall fixtures. Dimmer controls fixtures.

Even though the generator is not operating, it is illegal to connect both power sources to supply power to the service-entrance equipment without providing a transfer switch. If the standby generator is operating without a trans-fer switch to separate the two power sources, power will be fed to the utility's lines with the possibility of electrocuting a lineman who might be working on the lines of the utility.

The auxiliary generator can be portable, to

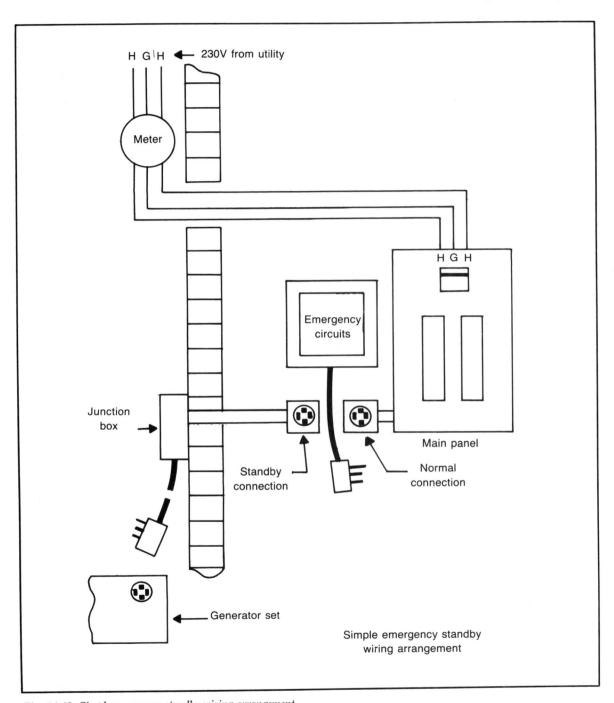

H G|H ← 230V from utility

Meter

H G H

Emergency circuits

Junction box →

Standby connection

Normal connection

Main panel

Generator set →

Simple emergency standby wiring arrangement

Fig. 14-42. Simple emergency standby wiring arrangement.

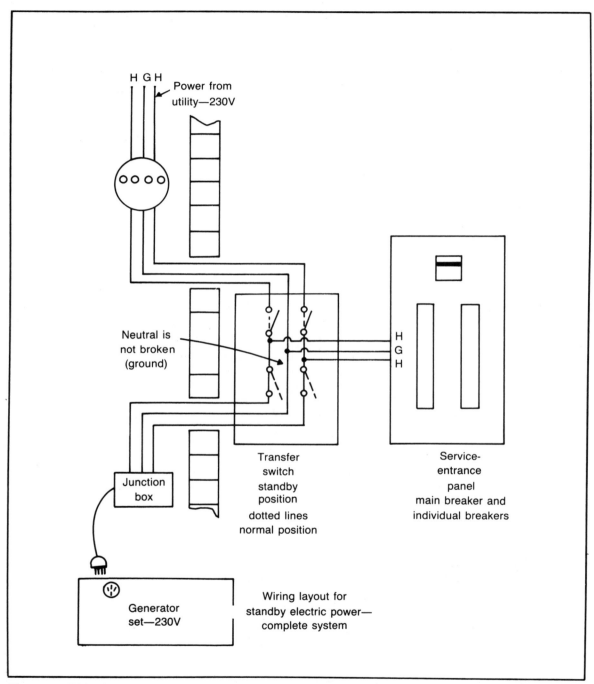

H G H

Power from
utility—230V

Neutral is
not broken
(ground)

Transfer
switch
standby
position

dotted lines
normal position

H
G
H

Service-
entrance
panel
main breaker and
individual breakers

Junction
box

Generator
set—230V

Wiring layout for
standby electric power—
complete system

Fig. 14-43. Wiring layout for standby electrical power for the complete system.

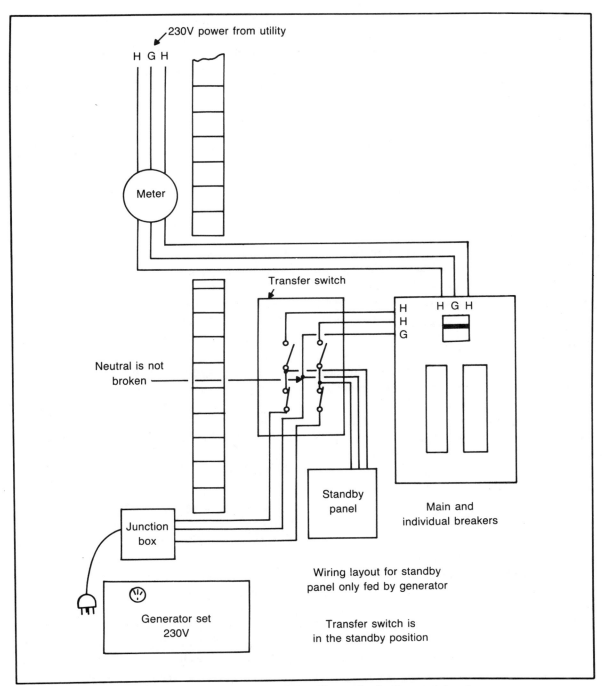

Fig. 14-44. Wiring layout for standby electrical power to the standby panel only, for essential needs.

be wheeled outdoors when needed, and the system can be plugged into the generator receptacle by means of a flexible cord. If the generator is to be used inside a building, the exhaust must be piped to the outside to eliminate any chance of carbon monoxide seepage into the building during its operation.

If the generator is to be permanently connected, it can be either inside or outside. An outside generator needs protection from vandalism and theft. This should take the form of a cage and perhaps lighting.

If a small-capacity system is installed, only certain loads may be connected to the system. The connected load might be the heating plant, refrigerator/freezer, a few lights, and a sump pump or well pump. These few circuits should be fed from a separate distribution panel to make it easy to connect as a separate entity directly (through the transfer switch, of course) to the standby generator. *Caution:* In any event, no auxiliary power may be connected to wiring supplied with power from a utility unless connected through a transfer switch.

If the generator set is to be permanently installed, a concrete pad should be made. This pad should be 4 inches larger all around than the generator base. This allows an area for the concrete anchors to be installed without the concrete cracking near its edges. Securely fastening down the generator in this manner helps prevent theft.

The connection box must be of weatherproof construction because it is to be mounted outside. The connection plug should be four-wire 240 Vac if your generator is to supply 240V. If the generator supplies 120 Vac, the plug will be three-wire 120 Vac. Be sure the assembly is of weatherproof construction. All wiring (including conduit and Thinwall) requires standard construction methods. You might need

to rent a bender to bend the Thinwall. Long-sweep elbows for rigid conduit can be purchased; three, at the most, might be needed. These elbows are connected to conduit lengths by means of conduit couplings. You also need LB condulets, offset nipples, straight nipples, and other common fittings.

Small-capacity generating systems are arranged for pull-starting similar to a lawn mower or garden tractor. These small systems are strictly manual and cannot be retrofitted to operate automatically. Because these systems are not designed for heavy loads, a simple manual changeover arrangement can be installed. This method requires two 50A three-pole, four-wire electric range cord sets. Each set consists of a cord with two No. 6 and two No. 8 conductors. These cord sets come in 3-, 4-, and 6-foot lengths. In addition, you need two 50A three-pole, four-wire range receptacles (outlets). This small, simple system will handle only essential needs. You have to decide what loads are absolutely necessary. These loads are:

▶ the heating system
▶ in rural areas, the well pump and related accessories
▶ the refrigerator/freezer

If a "life support" system is to be operated from this standby generator system, then its connected load must also be considered when calculating the total load connected to the generator. A 25 percent excess capacity should be added for safety.

Assemble the installation in the following manner. Mount a four-circuit breaker/fuse panel adjacent the main distribution panel. Do not connect this four-circuit panel to the main panel. Install next to the main panel one of the 50A range receptacles in an appropriate metal box.

Connect this box to the main panel by a 1¼-inch nipple, locknuts, and bushings. You might be able to get an offset nipple to make an easier connection.

Using No. 6 stranded copper wire, connect the appropriate terminals on the receptacle to the bottom ends of the main panel bus bars. These will have terminals for this purpose. A 3-foot piece of wire should be long enough for the connection.

Attach the terminal end of the range cord to the line terminals of the new auxiliary panel (this is the standby panel). You will connect all the essential equipment to this panel. Check the total connected load, because where there are four circuits, there should be enough to have each essential item on its own circuit. Be sure to provide for a few lights on one of the circuits.

In all instances, the fourth terminal should be the ground. This range cord plug now plugs into the range receptacle you have installed and connected to the main panel bus bars. If it is more convenient, you can turn the four-circuit panel upside down or sideways to accommodate the range cord so that it can plug into the receptacle without kinking.

The second range receptacle should now be installed near the first receptacle. Make sure the range cord plug can be inserted into either receptacle easily. This second receptacle is to be wired to the generator junction box with conduit. You can use four-wire Romex, size No. 6, between the receptacle and the junction box. This junction box then has the second range cord connected to the ends of the wires from the "standby" receptacle at the standby panel location. The plug of this cord then plugs into the generator receptacle that is on the generator panel.

If the generator is to operate inside the dwelling (as in the garage), the exhaust pipe must lead the fumes outside and away from the building. This is *very* important! Also, if the generator set is to be housed indoors, two louvered openings must be provided for engine ventilation; one for air intake mounted near the ceiling and one for exhaust near the floor line.

In normal use, the range cord and plug dangling from the standby circuit panel plugs into the main panel receptacle. In an emergency (standby), this plug will plug into the emergency receptacle. Mark these two receptacles correctly so they can easily be seen and properly used. Depending on the nearness of the two receptacles to the cord plug, you might only need a 3-foot range cord. Refer to Fig. 14-42.

If the generator is to run on natural gas, note that the natural gas supply might be cut off in event of an emergency. If the generator is to supply power for life support systems, it would be wise to use LPG for engine power. The use of gasoline is not recommended because of the high flammability of the storage facility and the possibility of leaks dripping on the floor. Both natural gas and LPG have odor detectors added to the supply for lead detection. If you decide to install such a system, contact various manufacturers to help you decide which one is best for your needs.

Chapter 15

Rewiring a Dwelling During Renovation

With the blossoming of many urban renewal projects, particularly in the inner-city areas, there is a great need for the know-how to do these jobs correctly and in a workmanlike manner. This is an important part of the "do it yourself and save money" theme of this book. Most of the inner-city urban renewal projects involve very old and run-down buildings.

It is easier to work on this type of building if the interior plastered walls are stripped down to the studs. This serves a dual purpose: it allows blanket insulation to be installed in the exterior walls, in addition to making it easier to install wiring, plumbing, and heating equipment.

At the same time, the other trades could install their work with less expense and time involved on their part. In other words, this would be new work and would be installed exactly as in a new building under construction. Generally in buildings this old, the plaster is in poor condition. New drywall would be installed

and taped after all concealed work had been done. The old plaster on the ceiling would also be removed to allow for wiring in that area. Friends of my parents did this very thing. They bought an old house and gutted the inside and started over as if it were a new building. Note also that after the plaster had been removed, the studs and joists can be inspected for defects or broken framing.

In addition, installation of plumbing, heating, air conditioning and of course the wiring progress much faster. Don't forget to install telephone and television wiring. Homeowners are now permitted to install wiring for their own phone system. Hence, if the original plaster is to remain, there will be much additional labor.

The removal of wiring can be started at any point. Consider salvaging attractive and antique lighting fixtures. These old fixtures can be refurbished and rewired to become very attractive additions to the renovated dwelling. Also usable and attractive are old metal (brass) switch and receptacle plates. I remember some that were called *oxidized copper*, a mottled antique copper finish. Push-button switches can be salvaged, but they must be in top condition; make sure that the contacts are not burned or pitted. Look closely, using a flashlight, to inspect the moveable contacts. Slight pitting is not that important, but the porcelain body must not be cracked (porcelain *can* stand high heat).

With all of the materials worth salvaging removed, you can now proceed with the removal or abandonment of the rest of the wiring.

REMOVAL OF ALL OLD WIRING

If the original plaster is to stay, as much of the wiring as possible must be pulled from the interior of the walls. Try to reach inside the wall cavity and cut loose as much wiring as possible. The main purpose is to eliminate any contact or even closeness of the new to the old wires. Take no chances that any of the old wir-

Fig. 15-2. An old installation of BX and Thinwall. The pigtail splices have been taped with rubber tape and friction tape. The wire connections may have been soldered as required. This is not known.

ing can ever come in contact with the new wiring. See Figs. 15-1 through 15-3.

After all of the old wiring is removed, the building is now ready for rewiring. If the plaster is in place, use old work installation methods. It is necessary to cut many holes in order to run the cable properly.

On outside walls where there are eaves, notches have to be cut at the ceiling-to-wall junction to bring cable from the attic down this wall. Use this method on one-story buildings and for the second story on two-story buildings. Gable ends do not need this treatment, as there is available room for working. Find methods of running the cable so as to minimize cutting and patching. This might mean using more cable to

Fig. 15-1. Remnants of a knob-and-tube installation. Top: a cleat that can hold two insulated wires. Below: a knob, which usually holds one wire with a stub piece opposite, or can hold a branch wire spliced near the knob and leading elsewhere.

Fig. 15-3. A fairly neat BX installation.

take a roundabout route to achieve the result.

Spend some time laying out the route to make sure you can go that way. Think before you cut and measure before you cut. Measure twice and cut once. Usually, first floor receptacles and switches can be fed from the basement or crawl space. First-floor ceiling outlets for a fixture might need to have the flooring cut in second-story rooms to run cable from the attic or basement. Try to cut flooring in halls or closets, if possible.

If the old plaster is completely removed, then the wiring method will be conducted as new work. This type of wiring work is straightforward and will go much faster. Therefore, I am in favor of removing the plaster. Very old plaster tends to crumble and does not take wallpaper or paint well. The speed and economy of new work over old work might pay for most of the drywall installation. You can remove the plaster and wiring, plumbing, and heating yourself. This work is very hard and dirty, so wear a dust mask and goggles.

REMOVAL OF OLD SERVICE EQUIPMENT

Sometimes the meter and service equipment is so old and primitive it can hardly be called "equipment." It might be only an open knife switch with two fuses for 120V. If you are doing a complete renovation, there should be no power in the building. You might need to arrange to have temporary power furnished or buy it from a neighbor. Always buy from neighbors; never borrow. If you hire other workmen, they need power also, so make arrangements.

With the power off, the service equipment can be removed. Tear it all out. If you want to save anything for nostalgic reasons, do so, but do not reuse these items.

Before installing any new service-entrance equipment, talk with the utility and get their recommendation for the meter location. The utility might not connect their lines to a location on the building if it has not been approved. The utility needs a direct route for their overhead lines. The original connection point to the building is usually satisfactory to both you and the utility.

The service-entrance panel in the basement or first floor must be immediately inside of where the entrance cable or conduit enters through the wall. Because this cable or wire is not fused or protected from overcurrents until it is connected to the main breaker or fuses, it must be as short as possible. You must make arrangements for space to mount the main disconnect and distribution panel or the combination main/breaker panel. This location must be adjacent the entrance of the cable or conduit from the meter. Be sure you have room for the equipment you choose. Measure carefully.

Combination main/branch circuit panelboards usually need less space. Even though they are more expensive, they fit in a small area. If the original panel was some distance from the cable entrance point, you want to install a main breaker panel there *only* and continue on to the distribution panel location with cable or conduit.

In my parents' home, the meter was in an upstairs bedroom and included fuses. These "fuses" were wire made of lead and secured under screw terminals. Many meters were located upstairs, because the wires from the utility pole entered the house at a high point. Again, fuse the service as soon as it enters the building.

Remove or abandon all wiring such as obsolete service-entrance wiring. This wiring might have been only size No. 10 because the usual service switch was only 30A. There were only two wires from the pole to the building, and therefore only 115V was available. Newer buildings had ¾-inch conduit on the outside of the building from the point of attachment from the pole down and into the basement by means of a gooseneck bend to make the 90-degree turn through the wall.

Present conduit size is 1¼-inch conduit or Thinwall. This will accommodate No. 3 wire to carry 150A ampacity. The ¾-inch conduit must be replaced by the 1¼-inch conduit/Thinwall or cable. Conduit looks nice especially when used with a mast and service head at its top.

Below the meter where the conduit enters the wall, use an angle fitting (known as an LB for *angle* fitting with *back* opening). This has a waterproof cover with a gasket and makes a neat job. Most fittings made for this purpose are offset back against the house wall to allow the conduit to lie flat against this wall. As my son says, this makes it "more better."

Condulet bodies are designated by location of openings. Example: type LB is an elbow body having a "back" opening and an "end" opening. Therefore it is an elbow (L) with a back

(B) opening. The cover is on the front with a rubber gasket to make it weatherproof. LL has the opening on the left: LR has the opening on the right.

Remove any service-entrance conduit. Cut it with a hacksaw at the gooseneck. It is best to get help when taking down conduit. They are heavier than they appear and can cause accidents. Long pipes or planks can be overbalanced when cut free.

Work carefully after removing the vertical conduit. Go inside and disconnect the conduit connection at the service-entrance equipment. After you have done this, the rest of the gooseneck and straight part will pull through the wall easily. Plan the layout for the new service. Thinwall (conduit) looks best for the part from the service head down to the meter base. You can use cable from the meter base all the way inside to the service equipment.

Cable is easier to work around bends and through holes in the outer wall. To keep out moisture, caulk the opening where the cable goes through the building wall. Fill this area well and then cover it with a *sill plate*. This plate is formed to cover the cable as it turns into the wall and is held against the wall by wood screws. Push the plate into the caulking compound and fasten it with screws. The caulking under the plate should be sufficient; if not, fill it in with more.

Be sure the cable is not bent at too sharp an angle. Cut away the top edge of the hole through the wall to allow the cable to make a long sweep curve. The minimum radius allowed is five times the diameter. Therefore, a 1-inch cable should not be bent in a radius less than 5 inches. Trim the hole edge accordingly. Sharper bends damage the covering.

Apply at the utility for the meter base. If you use conduit above the meter only, assemble the meter base, conduit and service head as one piece. Some bungalow and ranch houses with low hanging eaves need to have the conduit go up through the roof at the point where the roof boards meet the wall. The mast (the extension of the conduit above the roof surface) must conform to certain code requirements as follows:

▶ The service drop connection at the service mast must have a minimum of 3 feet clearance from the roof surface, provided the roof has a pitch (slope) of 4 inches in 12 inches. The connection of the service drop to the mast must be below the service head to prevent moisture from entering the service head.

▶ The service drop connection may be a minimum of 18 inches if the conductor cable passes over no more than 48 inches of roof surface. This refers to a 48-inch roof overhang (eaves).

▶ The service drop cable must maintain a minimum of 10 feet above the ground at any point between the building and the connection at the utility pole.

All of these conditions refer to cabled service drops having two insulated wires, the grounded messenger (support) cable (used as a neutral), and they are limited to 150V to ground. This applies to residential service only. If the service mast must extend some distance above the roof, the mast should be 2-inch pipe and it might need guy wires to take the strain of a long service drop span.

Winter snow and ice conditions, combined with strong winds, put a severe strain on the service drop, and in turn on the mast. Guying the mast is important for these reasons. The 2-inch pipe conduit used as the mast needs to be bushed down at the meter base. The hub is usually 1¼-inch female thread.

Many old two-story buildings have a service drop consisting of three separate (or two separate) wires supported by a metal rack holding three porcelain insulators separated from each other by about 6 inches. Sometimes the wires are attached to individual insulators screwed directly into the building framing by means of lag screws cemented into the insulator base. Check with the utility to find out if they want to run a new cable service drop rather than reuse the separate wires. The cable drop is stronger and causes fewer problems.

Beginning in the attic, cut and remove all the old wiring you can see. Where wires go down through the walls, try to pull them out. Wear gloves so that you do not cut your hands. The old insulation can break off and leave bare No. 14 wire that, when pulled through bare hands can give a deep cut. You need your hands in good shape to finish the job, so be careful.

On the floor below the attic, remove all electrical devices in walls and ceilings. After removing these devices, pull all the wires from inside the walls. Again, protect your hands for the same reasons. Any wires that cannot be pulled loose and removed must be cut off well inside the walls. Remove all the old wiring, and then go to the basement and do the same there.

With all the old wiring removed, lay out the new wiring plan. You should have a tentative wiring plan on paper, as required by the inspector. Remember, no spot along any wall may be farther than 6 feet from a receptacle. Provide for three-way switching where needed. Three-way control is necessary in hallways, stairways, and for large rooms such as family rooms and living rooms having more than one entrance to the room. Rooms such as bedrooms, living rooms, and dens usually have no ceiling fixture. See Fig. 15-4.

Consider installing ceiling fixtures. You can provide for this by wiring for the ceiling fixture and covering the ceiling box with an attractive ceiling canopy cover made for this purpose. This is a decorative cover held by a knurled brass cap nut. This forethought will eliminate running wiring later, after the ceiling has been decorated.

I installed fixtures in both of my bedrooms (I bought the house after completion) and find

Fig. 15-4. How not to do wiring. This installation would not pass inspection.

it very convenient. I removed the switch control from the wall receptacle and now control the ceiling fixture from this switch. This is approved by the code.

Living areas need switch control of receptacles so that lamps that are plugged into them can be controlled from a wall switch when entering or leaving the room. Many contractors ''split-wire'' these receptacles so that the upper outlet is hot at all times and the lower one is controlled by the wall switch. The reason the upper outlet is left hot is for appliances such as a vacuum cleaner or electronic equipment. A clock radio must be energized continuously.

Wiring for stereo, TV antennas, and telephone service is low-voltage or no-voltage work, but you must be careful to make proper connections just as with line-voltage wiring. While there is absolutely no fire hazard, work accurately so that you will not have to tear a wall apart to repair defects. Check all circuits and route all wires to prevent damage to these wires. Use the same construction procedures as when working with the line voltage wiring.

Wall boxes might not be necessary except for protection (because the NEC does not apply to this class of wiring). The NEC does prohibit this class of wiring being in the same cabinet or wall box as line voltage wiring because this wire does not have enough insulation for line voltage. The only allowable situation is where there is a so-called metal partition (barrier) in the box separating these two classes of wiring systems.

OLD WORK WIRING METHODS

Old work wiring is different from new work because in old work, cables must be drawn through inside the walls without doing extensive damage. In addition, the walls must be patched where access openings have been made. To do work that is acceptable, the patching must be practically invisible. By careful cutting, the patching is made easier and less noticeable.

Cable can be run behind baseboard for a horizontal run between two receptacles where no other route is possible. To do this carefully, remove the baseboard by prying slowly with a wide wood chisel driven behind the baseboard. Pry outward gently along the full length, a little at a time. The modern thin baseboard tends to split or crack, so be extra careful.

Do not drive the nails back through the face of the board. Use your electrician's side cutters to grasp the shank of the nail close up to the back of the board. The point of the nail is facing you. Now pry up using the jaws of the pliers as a fulcrum. If the nail is a finishing nail, it will pull right through the wood. If the nail has a flat head, you should drive this type back through the front of the baseboard. One pry should pull the finishing nail through; if not, grasp and pry again.

Caution: In any type of remodeling work, always remove all protruding nails. It is extremely dangerous to step on a nail. On lumber to be discarded, the nails should be bent over and the lumber discarded. Just be sure the nails do not stick up. After removing the baseboard, you will find the bottom 2 × 4 (plate) and the bottom ends of the studs nailed to the plate.

If the baseboard is high, say 4 or 5 inches, you should be able to cut away the plaster to within 1 inch of the top edge of where the baseboard top edge covers when it is replaced. This method will allow you to drill holes through the part of the studs showing behind where the baseboard will cover. You might need to use a drill extension (available in hardware stores) to

use an electric drill motor. Drill in the exact center of these studs. This is an NEC requirement.

If the baseboard is narrow, you have to notch the stud or the base plate. The very bottom of the studs should be behind the baseboard. Cover the cable with a ¹⁄₁₆-inch metal plate to protect the cable from nails. Do this to protect yourself and prevent the house from burning down. It might be necessary to recess these plates to allow the baseboard to fit back in place. Do not replace the baseboard until inspection has been made.

When you bring cable down from the attic on an outside wall, there will be places where low roof lines do not allow access to the top plate, so you have to cut a notch in the interior wall and ceiling corner. To do this, cut a notch half in the ceiling and half in the wall to make an area exposing the top plate. The notch should be about 1½ inches wide by 1½ inches deep, both on the wall and on the ceiling. It will be necessary to notch the plate and then cover it with ¹⁄₁₆-inch metal.

Installing Wall Boxes in Lath and Plaster

When cutting lath and plaster for wall boxes, center the box vertically on one lath, even if it means raising or lowering the box location slightly. If you do not do this, it will be more difficult to install the box. To locate the center lath, probe with a nail or screwdriver to locate the top and bottom edges of the lath. Mark the box outline on the wall.

Carefully remove the plaster for the full width of the box horizontally from the lath and by the full width of the lath vertically plus the space between this center lath and the edges of the laths above and below the center lath. This gives a plaster-free space about 2 × 2 inches. Refer to Fig. 9-20.

Now carefully remove the plaster from the box outline top and bottom of the plaster-free opening. Note that the center lath part showing will be removed. Allow a slight amount of slop (short for sloppy fit) when removing the plaster. Try to place the front of the box into the plaster-free area to see if it fits. This opening should be ⅛ of an inch larger, in both dimensions, than the box dimensions.

Starting with the center lath, saw down one side of the lath, next to the plaster edge, with a coarse tooth hacksaw blade (tape one end of the blade to make a handle) or a keyhole saw. *Caution:* Do not saw either side of the lath all the way through; saw about half way on each side. Insert a wide blade screwdriver in either side and twist the blade so the sawed portion splits off. If you have no luck, try the other side. If the grain is slanting, you might have to saw more on one side for the piece to split off. Also saw up from the bottom edge and split off this piece.

Now hook one finger behind the remaining center piece to support it while you saw through both ends. This is to prevent the sawing from loosening the lath from the plaster "keys" (the plaster that squeezes through between the laths and locks the plaster to the laths). Saw gently.

Now do the same with the part laths for about half the width. This will provide the full opening for wall box. Be sure to hold these laths to prevent breaking the plaster key. You have to provide a notch in these half laths to allow the device mounting screw to clear them. Make a small notch in the lath edge with the hacksaw blade. If the wall box will not fit, you might have to pare the edge of the lath with a pocketknife.

There are two methods of using the mounting ears furnished with the box (plastic boxes usually have permanently mounted ears). Metal boxes are built so that the ears are reversible.

With the ears flush to the box front edge, they will be on the surface of the plaster. With the ears reversed, the plaster will have to be removed and the ears screwed directly to the wood lath. This is the better method, because it is easy to crack the plaster and you will have to remove it anyway.

In either case, use No. 6 or 8 flathead screws. Do not remove too much plaster or the cover plate will not cover the space. You have to plaster around the box to comply with the NEC requirement. Any trouble encountered with the wood laths can be overcome by using the Madison supports or a box with a clamp arrangement built in. When surface-mounting box ears, you should remove some plaster to allow the ears to lie flush with the wall surface.

Where ceiling boxes are to be installed in rooms, find the room center. (See Chapter 9.) If you run into a joist at the exact center, move the box slightly to one side, but more than 1 inch will be noticeable. You can use a shallow, ½-inch box and notch the joist a small amount if necessary. Secure the box to the joist with No. 10 ¾-inch wood screws or sheet-metal screws. Do not use nails because you might have to remove the box for some reason. When the room center falls between joists, use the standard adjustable bar hanger, made from 11½-inch to 26½-inch adjustment (two sizes). Separate the two halves and reverse the stud that holds the box in place on the bar.

If you can gain access to the attic or have cut access openings in the second story flooring, this hanger can be nailed to the joists on each side of the box location. If this is the case, use a standard 1½-inch deep box to allow more wiring space.

If the top side of the ceiling is not accessible, use the ½-inch deep box and let the bar hanger lie on the top (back) side of the ceiling finish (drywall or lath and plaster). The other alternative is to use the spring out clamp shown in Figs. 9-6 through 9-9. This type works with a ceiling box to support a lightweight fixture only. Don't forget to plaster around the box after installation.

All concealed wiring, such as rewiring an old house, must have the new wiring in place before the boxes are installed. This is just the opposite of new work where all the boxes are mounted before the wiring is run. Cut all openings and drill framing members before pulling any cable.

You should now have replaced the old service-entrance equipment with new and be ready to have the utility connect their lines to the house. This way you can use power tools and save time. After you have finished the entrance equipment and have had it inspected, call the utility for the connection. You should already have done any outside remodeling of the building (aluminum siding or face brick) before mounting the service mast and meter base. Check with both the inspector and the utility before proceeding too far with the service-entrance equipment installation. Remember to do the work only yourself (family help is OK). The hiring of outside contractors is prohibited by local ordinances.

Review Chapter 4 and Chapter 5. These chapters give detailed instructions and advice for working with these organizations. Also refer to Chapter 13. Figure 13-31 shows methods of reconnecting the separate circuits using service-entrance cable, and a new combination main and distribution panel. A junction box takes the place of the old service-entrance equipment. Figure 13-39 details the replacement of old ¾-inch service-entrance conduit with a new service mast and outside meter. A third wire will need to be added, if it is not already in place,

to provide 240V service for a range, air conditioning, and water heater. Figure 13-33 shows procedures to remove or abandon old wiring in a building being remodeled.

Details of drilling for cable from the basement to the attic are shown in Fig. 13-35. Figure 13-36 shows the arrangement of a receptacle (hot all the time) and a switch and ceiling outlet controlled by the switch. The source is from the basement.

Chapter 9 shows other old work methods. Figure 9-14 shows the use of a mouse (weight) and the method of hooking the cord holding the mouse from inside the hollow wall. Also shown are box supports. These supports can also be used for mounting wall fixture boxes. Figure 9-19 shows a method of removing and replacing finished flooring to gain access to a ceiling fixture location on the floor below.

Figure 13-31 details how to install a new service-entrance panel and new fuse/breaker panel. If the new panel is some distance from the old fuse panel, replace the old fuse panel with a junction box. Some of the circuits leaving the panel might be too short (the cables might be too short to connect to the new terminals). Note that the new equipment can be a combination main/distribution panel or each part can be separate.

In certain areas, Hawaii for example, the meter and main breaker are mounted outside in a weatherproof enclosure. The wiring between the main breaker and the distribution panel is four-wire service-entrance cable, using an insulated neutral wire rather than the common bare neutral in standard service-entrance cable.

In this situation, the NEC requires this type of cable. If the new junction box is within 2 or 3 feet of the new distribution panel, you can use conduit or Thinwall between the two enclosures. Usually, though, four wire entrance cable is better and more convenient to work with. It might be difficult to purchase short lengths of insulated wire in large sizes of No. 6 or No. 4.

When using Thinwall or conduit, buy all black wire and tape the neutral with white adhesive or plastic tape to designate the neutral. Tape each end of the neutral. You should identify the two hot wires from the main breaker. Put one band of black tape on the left hot wire. Put two bands of black tape on the right hot wire. The "left" and "right" are when looking at the main breaker terminals. The four-wire service-entrance cable comes color coded: black, red, white, and bare. The bare wire is the same as that in Romex or BX and is used to ground the cabinets and other noncurrent-carrying metal parts of the system.

When using conduit or Thinwall, tighten all locknuts and bushings well to provide a continuous ground path for fault current anywhere in the system.

The grounding of any electrical system is extremely important for personal safety as well as providing for prompt blowing of fuses and breaker operation when necessary. In case of a poor grounding system or no system at all, the overcurrent protection might not work and could cause a fire. The best ground is that provided by a cold-water pipe connected to the city water supply. This is the common grounding connection in urban dwellings.

This ground connection is attached to the neutral bar in the service-entrance panel. The neutral bar is the white multiple terminal connector bar where the white wires from Romex and BX cables terminate (connect). The neutral from the service-entrance cable connects to the bar, usually at the top end, close to where the cable enters the cabinet. This terminal is larger than the branch circuit connections. The water meter must have what is called a jumper wire to bypass the meter space in case the meter is removed for any reason. Some meters are connected to the water pipes by unions that have rubber gaskets. This provides a very poor ground connection, therefore the jumper is needed for this reason also.

Review Chapter 8 for a better understanding of grounding principles. This chapter stresses many important points, such as the continuity of the ground path from the furthest receptacle or switch all the way to the actual connection to ground (meaning the earth). The ground wire from the service-entrance equipment must be continuous all the way to its connection to the cold water pipe.

Buy the best-quality wiring materials and devices you can afford. Use only copper wire. Do not use aluminum wire. Follow approved wiring methods and practices. Make all electrical connections tight. When running Romex or BX, leave 8 extra inches from the wall and ceiling boxes for making connections. The NEC requires only 6 inches, but sometimes a little longer is better.

Notice that each box has the cubic inch volume stamped or embossed on the inside. Section 370-6 of the National Electrical Code specifies the number of conductors of a specified size allowed in each size box. Tables 370-6(a) and 370-6(b) list these requirements. I advise that you obtain a copy of the NEC in complete form for about $22.00. Calculation examples for determining the connection load for a single dwelling appear at the end of the book.

Caution: Some areas will not issue a homeowner's electrical permit to wire a two-family dwelling. Check this requirement before doing any work. You might be prevented from doing wiring for the unit you do not occupy. For calculating the connected load, use the code Example No. 1(c). Chapter 5 gives an example that is already worked out. Also review Chapter 7.

Remember to ask the inspector about anything you don't understand. Most inspectors are interested in having you take out a permit because they want to see that you are doing the work properly and making an electrically safe installation.

Additional Items to Consider Installing

Consider installing outdoor perimeter lighting to be controlled by a photocell. The photocell method is the best method because it requires no adjustment for length of day or night. The only requirement is that the photocell must be installed where it can sense darkness, but is not affected by lights such as car headlights, street lights or lights on the building next door.

Photocells are usually mounted at least 8 feet above ground level to prevent vandalism. Use an approved installation method. Figures 14-19 and 14-20 illustrates photocells to control lights on the exterior of a home. These can be floodlights and decorative fixtures. A metal box cover is used to support the photocell.

If you expect to heat domestic hot water with electricity, consider installing a water heater timer. Timers can save 20 percent or more of the electrical consumption needed to heat water. Certain models of timers have additional trip levers to allow more than one operation (on and off cycle) in 24 hours. If no one is home during the day, the heater can be turned off for a daytime period and also during the night to achieve greater savings. Install it yourself and save money and electricity.

If you have an attached garage or expect to add one to the dwelling, you should install a garage door operator. The operator will require a separate fused/breaker circuit with no other load connected to it except the operator. For the connection, install a receptacle in the garage ceiling about 24 inches beyond the top edge of the door (when it is in the open position and centered between the door tracks). This location should be within reach of the attachment cord of the opener motor. Chapter 14 suggests other projects that you can incorporate into your wiring plans. Reading about these projects will give you ideas for your own pet projects. The instructions given will usually apply to any additional work you will do.

(As a suggestion not strictly electrical, make arrangements to have all meters—gas, water, and electrical—located outside so the meter readers do not have to enter the premises. You can move the electric meter outside yourself. The gas meter usually has to be moved by the gas utility, for a fee. The water meter can be retrofitted with a remote reading indicator, which is a small plastic case mounted outside on the building, to be read from outside.)

Make complete detailed plans to include all special wiring you feel you want now or will need at a future time. Electric clothes dryer, air conditioning, and electric water heating circuits should either be installed (roughed in) or provision made in calculating the load requirements so that the service-entrance equipment can be sized properly for later installation of these fixed appliances.

Electric resistance space heating is very costly to operate. It should not be used unless there is no other alternative. Even though the installation cost is the lowest of all types of space heating, the operating cost is prohibitive. The low installation cost lulls people into having this type of heating installed. When the monthly bills start coming, the homeowner is usually extremely disappointed.

HOME SECURITY SYSTEMS

Because of break-ins, burglaries, and vandalism, there is much concern about protection of property. Homeowners are pursuing methods of discouraging such actions by means of outside lighting and varying lighting patterns inside the house to give the impression of occupancy when the premises are not occupied, as when you are on vacation or other reasons. Such systems can be programmed to operate various appliances as well as control heating and air conditioning.

Typical home control system applications can be separated into four categories

▶ convenience
▶ security
▶ energy savings
▶ cost effectiveness

A representative system has been developed by Leviton Manufacturing Co., Inc., Little Neck, New York. This system operates in all four areas. Because special wiring is not needed, costs are reduced over other systems on the market. Modules consist of devices that replace receptacles and switches. In addition, a wall-mounted controller can be used to turn on or off all outside lighting or operate other functions of the system.

The brain of the system is a wall-mounted programmer. One excellent feature is that the program signals are sent over the present electrical wiring by a special signal imposed on the wiring. All modules respond to their signal only and perform their functions. A wall-mounted controller can be installed in the master bedroom or in other rooms as well.

The wall-mounted programmer functions range from turning on the coffee maker (or the Christmas lights) to turning off the lights in a detached garage (no extra wires to run to the garage either), to night setback of heating equipment or night operations of air conditioning.

MINIMUM REQUIREMENTS FOR EXISTING DWELLING UNITS

Minimum standards for existing dwelling units regarding service-entrance equipment require 100A service equipment having three-wire capacity, dead front (no live parts exposed) and type S fuses (time delay, nontamperable).

There is an exception to this rule that provides for the continued use of 55A service equipment having three-wire capacity and feeders of 30A or larger. Two- or three-wire capacity is accepted if adequate for the load served. The existing wiring must be in good repair. Evidence of inadequacy is noted in any of the following:

(a) Use of cords in lieu of permanent wiring.
(b) Oversizing of overcurrent protection for circuits, feeders or service.
(c) Unapproved extensions to the wiring system in order to provide light, heat or power.
(d) Electrical overload.
(e) Misuse of electrical equipment.
(f) Lack of lighting fixtures in bathroom, laundry room, furnace room, stairway or basement.

These rules are excerpted from the *Reciprocal Electrical Council, Inc. Handbook.* Their address is 151 Martin Street, Birmingham, MI 48012. The amendments are adopted by the Department of Housing and Urban Development (HUD) Washington, D.C. available as *HUD-PDR-631-6*. The handbook is available from the Council for $1.50 plus postage and handling.

As the above rules specify, some of the present wiring may be reused. In very old buildings, however, the wiring is suspect and it should be carefully inspected. Any parts that do not appear in excellent condition should not be reused. Be cautious in this respect and make sure everything is safe.

INSTALLING TELEPHONE WIRING

It is now legally permissible to wire your own home for telephone service. Homes have

Fig. 15-6. Four terminal plates for connecting various phone jacks in the building.

a device called a Network Interface (NI). This NI is a connection point from which the interior wiring starts. The telephone company provides this NI, which is a modular jack, similar to those found in homes for plugging in a phone. See Figs. 15-5 and 15-6.

Telephone wire, jacks, long cords and related devices to wire your home are sold in hardwares and home centers. Jacks can be flush-mounted similar to electrical receptacles, or surface-mounted, with the cable run exposed. Some jacks are duplex and can be wired to provide for a telephone and an answering machine to be plugged in. In other cases, these two jacks can be used for two separate telephone lines (two different telephone numbers). Telephones are now available that are wired to two different and separate lines. Buttons on the telephone can select which line to answer or use, depending on which button is lighted, similar to those used in business offices.

Fig. 15-5. The electrical protector furnished by the phone company has been moved but has not been installed properly as nails have been used. Upper center terminal wire goes to a ground.

The telephone company may or may not inspect your home for customer-installed interior wiring, but the company might refuse to connect your interior wiring to their lines if it appears to be done incorrectly or poorly. The wiring must conform to Article 800 of the NEC, entitled "Communications Circuits."

Requirements are simple. One rule is that telephone wiring must be kept at least 2 inches from lighting and power wiring. Insulation on telephone wires must not support combustion. Article 800 of the NEC is concerned mainly with a neat, workmanlike installation, free of hazards to life and property in the use of such a system. This is similar to the rules for any wiring carrying less than 300 volts.

As telephone wiring consists of small size wiring (#22 and #24 AWG), it must be protected by the same methods used when installing electrical wiring. These methods include running it parallel to beams and joists, or through holes drilled in the center of these framing members.

Wall boxes are not required, but some means of support for the wall jack must be provided. One way to do this is to make a hole in the wall for the connection boss on the back side of the jack plate. This needs to be about 2 inches by 2 inches, then use toggle bolts or Molly anchors to fasten the plate to the wall surface. Running the wire from location to location is easily done when the house is under construction. When the telephone company did the "prewiring" of a new house, they arbitrarily left a small coil of the wire behind the plaster or drywall in various locations (one or more in each room). The NEC does not specify the number or location of these outlets as it does for electrical outlets.

The telephone company has a special sensing device to locate these wire coils. When a coil is located, the wall can be opened and a phone jack installed there. If you prewire your own home during construction, you should mark the location of these wire coils so you can locate them when you want to install a telephone jack. Do this carefully so that you can find a wire coil at the point where you want a telephone jack.

A better way is to install blank plates at the coil locations for ready identification. These plates show, but they do look neat. You can either install a plastic or metal electrical wall box and mount the blank plate on these, or install the blank plate using Molly anchors. With these anchors, the screws can be removed and replaced any time.

Telephone jacks cost from $1.75 to $2.50, depending on the brand. Those made by AT&T seem to be the best, but cost more. I have found other brands that will not make good contact unless the plug is pushed in as far as possible, even after the click. In most, when the click is heard, good contact is made.

Many modern telephones of whatever brand need only two wires to operate. The cord from the telephones to the wall jack have only two wires even though the jack is designed for four wires. It may be difficult to tell which two wires in the four-wire cable in the wall are the ones to use to connect to the jack. In this case you need to use a multimeter that reads voltage. The voltage supplied to telephone company equipment is 48 Vdc.

The cord connecting the telephone set to the wall jack now has only two wires. These are connected to the two center prongs on the plug. Now look at the back of the phone jack and determine which two wires connect to the two center terminals of the jack. The two remaining wires are not needed, although they can be used to connect to the other outlet of a two-outlet jack, if an additional separate telephone line is desired. Thus, two single-line (standard) telephones can be plugged in at this location. The telephone company will provide two separate lines if requested.

Many variations of equipment can be attached to telephone lines, such as standard one- or two-line telephones, answering machines, and telefax equipment. When buying equipment, buy top quality. Look for AT&T equipment on sale when you are ready to buy.

Small insulated staples, available in hardware stores, are used to support the cable in basements and attics. As mentioned before, follow joists and studs to protect the cable from damage. Drill holes through joists and studs when it is necessary to cross these framing members. The cable can be concealed within the hollow walls even when installed after the building is completed. When installing interior wiring in completed buildings, telephone com-

panies did not conceal their lines but stapled them to baseboards and around door or window trim. In commercial buildings, the electrical contractor installs conduit especially for the telephone wiring. This is the standard practice in the industry.

INSTALLING A FAN AND HEAT LAMP IN A BATHROOM

The addition of a fan and heat lamp is fairly easy. The present light or lights are controlled by a switch at the door. Turn off the power at the panel and remove the switch plate and mounting screws from the switch. Pull the switch out of the box so the wiring can be examined. You will find either of the following: either one or two cables enter the switch box.

If one cable enters the switch box, power comes to the light fixture box. Another cable also enters the fixture box. This is the switch

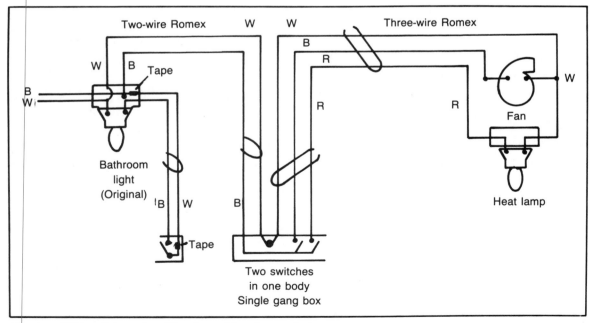

Fig. 15-7. Wiring diagram for adding fan and heat lamp to bathroom.

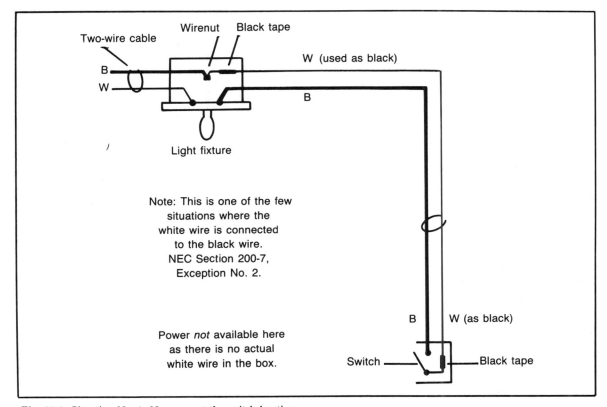

Fig. 15-8. Situation No. 1. No power at the switch location.

"leg." This cable runs from the fixture box to the switch box, as shown in Figs. 15-7 and 15-8. Loosen the fixture to find two or more cables. If the correct procedure was followed upon installation a white wire is connected to a black wire (from cable to cable), leaving two wires connected to the fixture, one black and one white. This arrangement means that there is no power (one black *and* one white) at the switch location. *Note:* Some people cheat and use the bare ground wire as the white wire (as in the switch box). This is illegal, in violation of the NEC, and could be *very dangerous*! *Note*: The connection of the white wire to the black wire in the fixture box *is* approved by the NEC

in this one situation *only*. The reason for this approval is that Romex cable is made with one black and one white wire and is not available with *two* black wires. It is recommended that the ends of this white wire in the junction boxes be identified as a black wire by wrapping a turn or two of black electrical tape near each end.

The other wiring arrangement has two cables coming into the switch box. See Fig. 15-9. One cable is from the power supply, and the other cable goes from the switch directly to the light fixture. The two white wires are connected together using a wire connector (wire nut). One black wire, either one, goes to one screw terminal on the switch. The other black wire goes

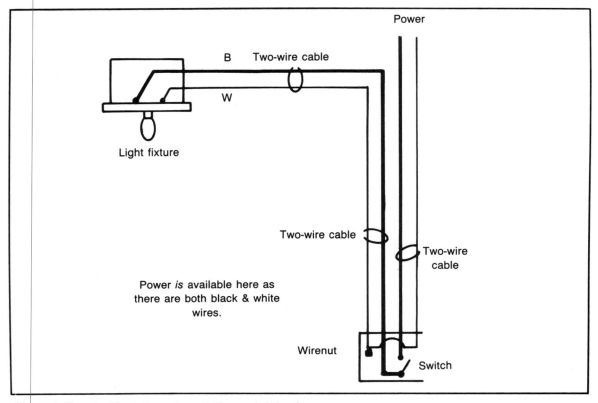

Fig. 15-9. Situation No. 2. Power is available at switch location.

to the other screw terminal. In this situation, you do have power at the switch location. Test for power by removing the wire connector from the white wires (sometimes you can just insert the prod of the tester into the bottom end of the wire nut without removing it from the connection). Touch the other prod to one of the switch terminals on the switch. With the switch off, only one of the terminals show power. With the switch on, both switch terminals show power. (You have to make this test with the power on, obviously.) Be careful! Work slowly and carefully. This power is then available to feed one or two switches for the fan or heat lamp or both. See Fig. 15-7.

WIRING A SOLID-WALL CEDAR HOME

The wiring of a cedar home is done differently than wiring a standard "double wall" home. In the cedar house, the wiring cannot be concealed in the hollow wall spaces, as there are none.

The areas for concealing the wiring are limited to the attic and crawl space or basement. Wall outlets are usually placed just above the baseboard and touching it. If the baseboard is thick enough, the back of it can be notched out to accept a Romex cable. Outlet boxes are surface-mounted. Wiremold makes such boxes that are just deep enough to take a receptacle or switch. You might be able to drill down at an angle, starting at a point just above the baseboard, into the crawl space or basement instead of notching the baseboard.

Cedar home manufacturers furnish a U-shaped molding that can be placed over Romex that was brought down from the attic to supply a switch at the normal 46- to 48-inch height. This molding is about ¾-inch thick and 1⅜-inch wide. The top surface has two parallel grooves ¼-inch in from each edge for decoration as well as a nailing place.

Other wires or cables such as telephone, thermostat and stereo wiring can be concealed in much the same manner. In certain situations, if a wall backs up to a closet, the cable can be run exposed on the closet wall (be sure to staple the exposed cable in the closet every 4½ feet), and then go through the wall to supply a receptacle or wall switch on the opposite side. As methods of concealment are few, one has to be resourceful when wiring this type of construction. The cables can be made less noticeable if they are painted the same color as cedar walls.

Use more cable if you need to take a roundabout way to feed a device or current-using appliance. It is best to plan ahead when doing this type of wiring. Mark locations for ceiling or wall fixtures, receptacles and switches. The living room, dining room, and bedroom areas are the most difficult areas to attempt concealment of the cables.

Wall receptacles should be back-to-back, such as living room and bedroom. Drill down from the bedroom to the crawl space or basement and also from the back of the receptacle box through the wall between the bedroom and living room. Thus, a short piece of Romex can be run through the wall to feed the living room receptacle. Do this in as many places as possi-

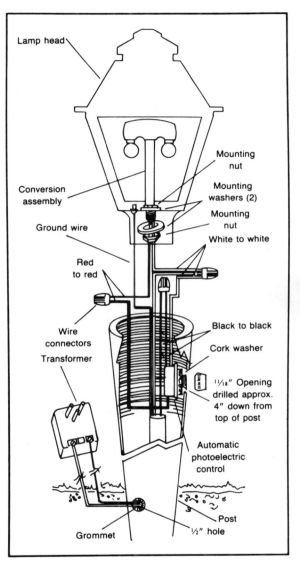

Fig. 15-10. Cutaway view of gas lamp. (Courtesy Harper-Wyman Co.)

ble and also with the wall switches on each side of the wall. When running cable for two wall switches (one on each side of the wall), use #14-3 Romex. This 3-wire cable (in this case) takes the place of two #14-2 cables.

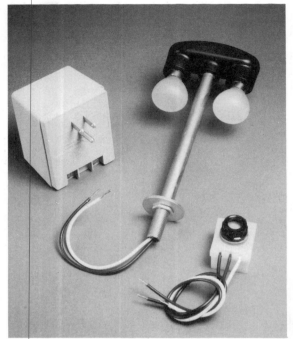

Fig. 15-11. Gas-to-electric conversion list. (Courtesy Harper-Wyman Co.)

gas in the basement or utility room that supplies only the lamp. *Note*: If this line has no valve, you have to shut off all the gas at the house main valve. Go to all gas appliances, namely, furnace, water heater, gas clothes dryer, gas stove and any others. Close all main valves to the OFF position. Make sure all gas stove burners except the pilot are OFF.

If you are lucky, there is a shut-off valve for the gas lamp. If so, close the valve and disconnect the gas line on the side of the valve going to the yard lamp. If the valve outlet is a flare, buy a flare cap at a hardware store that stocks flare fittings. The flare cap will be ⅜-inch or ½-inch in size. Move the end of the gas line to

Fig. 15-12. Complete lamp. (Courtesy Harper-Hyman Co.)

Referring to Fig. 15-13, this diagram shows the use of 14-3 w/ground Romex to provide power to a switch in the living room from a switch in the bedroom. Check the diagram carefully, and it will show you how to wire these switches properly. This eliminates using two lengths of 14-2 Romex. The bedroom is on the left in the drawing.

CONVERTING A GAS YARD LAMP TO ELECTRICITY

Because a gas-operated yard lamp burns 24 hours a day, it uses a great deal of natural gas. It is more economical to convert the lamp to electricity.

To make this conversion, first shut off the

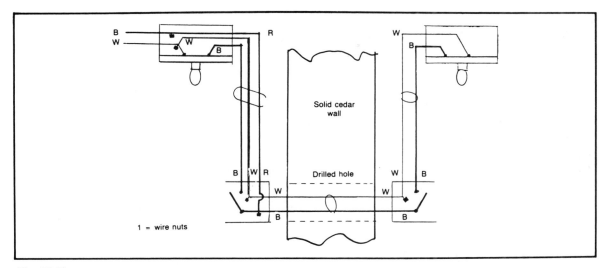

Fig. 15-13.

the yard lamp away to one side, then install the flare cap. Tighten it firmly. The gas line can be cut off where it goes through the basement wall, and the rest of the line that is underground to the yard lamp can be abandoned or dug up. If the valve outlet has a pipe thread, buy a pipe cap or pipe plug to close the valve opening. If there is no valve, there is usually a pipe union. Disconnect the union and remove the half of the union remaining on the supply end and cap it. Nearly always, the pipe supplying the yard lamp is ⅜-inch O.D. copper, but all sorts of installations are possible. You can now turn the gas back on at the main house valve. Light the pilot on the gas stove first, then the furnace, water heater and gas dryer. These three appliances generally have printed instructions on a plate on the front. Turn the knob to PILOT, then while holding the knob down, light the pilot. If the pilot does not stay lighted, repeat the procedure. If the pilot stays lighted after you release the knob, you can then turn the knob to MAIN VALVE.

The lantern assembly of the yard lamp will have to be removed. See Fig. 15-11. First remove the glass, then loosen any set screws that are holding the head. Remove the entire head. You should be able to remove the gas burner by unscrewing it. This should give enough room to run the Romex up inside the post. This post is usually 2 or 2½ inches diameter, which allows plenty of working room.

The best way to run the wire is to drill a ⅞-inch hole in the post, 6 to 8 inches above ground level, and install a ½-inch bushing (hex on one end and ½-inch threads on the other, with a smooth, round edge opening for the wires). Also get a ½-inch locknut. This next procedure is tricky. Refer to Fig. 15-10. Take the bushing and wrap about 6 inches of a small gauge solid wire about six feet long around the threads, close to the hex part. Lower this down inside the post so that the threaded part can be fished out the ⅞-inch hole in the post, using your handmade fishing hook. This takes two people, one to fish and one to hold the bushing once it is through the ⅞-inch hole in the post. Hold the bushing in place in this hole, using a 1 × 2 length of wood down the inside of the post. When you get the bushing threads through the hole, screw the locknut on lightly, then screw on an angle fitting with the other opening pointing down.

In the house basement, locate a lightly loaded or spare circuit to get power for your lamp. The procedure at the house end is similar to that at the post end. In a frame house, a hole is drilled in the wood trim above the foundation to take a ½-inch piece of Thinwall (EMT) conduit. The inner end of the Thinwall should connect to a junction box, preferably 4 inches square. The Thinwall can be brought into the back (bottom) of the box. Therefore, on the outside of the house, you install an angle fitting (½-inch type "LB" Condulet). The Thinwall will extend from the "LB" through the wall and enter the back of the 4-inch box. A Thinwall connector is needed on both ends of the Thinwall. Measure the distance through the wall and allow the proper extra length on both ends, about ¼ inch on each end. The inner 4-inch box serves as a junction point from the interior Romex to the underground cable.

After these above-ground fittings have been installed, you can now dig a trench for the underground cable. This cable must have the designation "direct burial," or similar statement on its covering, denoting it can be used for this purpose. You might want to dig up and remove the copper gas line and use this trench for the new electrical cable. Underground cable can be pulled through conduit or Thinwall for additional protection if desired, but this is not mandatory. Short sections of Thinwall extend down from the LB at the house and the angle fitting at the post. Then make a 90-degree bend, to lie parallel in the trench bottom. If the Thinwall ends in the trench and does not continue, this end must have a connector and bushing (an NEC requirement). For residential work, the trench must be at least 12 inches deep for protection. Eighteen to 24 inches deep is better, but not mandatory. It is recommended that a 1 × 6 board be laid over the cable in the trench as an extra precaution; the belief is that a person digging in the area would investigate upon finding wood, rather than cutting through the exposed cable with a shovel.

Run a length of Romex from a live circuit that is always on, such as a pull-chain basement light or wall outlet, to the new 4-inch box that you installed. As with any extension of a circuit, run the complete extension before making the final connection. The underground cable can be run up inside the post and to the connection at the lamp socket.

Most conversion kits now are low voltage (12 volts). These kits recommend that 14-2 UF (underground) cable with a ground be used. Included is a voltage plug-in transformer. When the cable is installed as shown in the directions, this method will not conform to the NEC. If the cable is installed according to the methods specified in the NEC, as described in the first part of this article, the system can use 120V bulbs. Thus the cable is wired to a live circuit and the transformer is not used.

Appendix

Procedures and Products

SOMETIMES IT IS EASIER TO SHOW SOMEONE how to do a job or procedure than to describe it. The information in this appendix shows how to install and wire a receptacle using a "gangable" metal wall box. The gangable wall box is made so both sides are removable. In this manner, individual boxes can be combined to provide room for more devices.

MOUNTING A RECEPTACLE IN A GANGABLE WALL BOX

The gangable box provides a view of the wiring for a duplex receptacle. These boxes have a screw on the top and bottom edge of the center section. When either of these screws is loosened, either side can be removed. By removing one side from each box, two boxes can be joined together to make one "two-gang" box. Tightening both screws secures the boxes. With both sides removed, additional gangs can

be added between the end boxes to make multiple-gang boxes.

Figure A-1 shows a receptacle that has just been wired to the Romex, supposedly coming into the box. The bare wire is the grounding wire. There are three of these wires showing: 1) from the Romex; 2) to the receptacle green grounding screw; and 3) to the box grounding clamp. This clamp is on the front edge of the removed side (in the lower left).

Figure A-2 shows the receptacle about to be pressed back into the box. All wires are attached (black to brass screw, white to white screw, and bare to green hex grounding screw on the receptacle). The grounding clip is on the box side near the top of that side.

Figure A-3 shows why the screw holding the receptacle to the box does not give secure grounding. Notice the space between the receptacle ears and the box ears (about $1/8$ of an inch). Because the screw is a loose fit in the recepta-

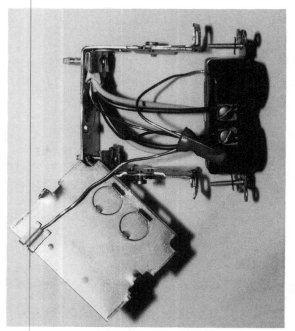

Fig. A-1. Wiring a receptacle in a gangable metal box.

Fig. A-3. Shows why grounding wire is needed when box is not flush with wall. Wires feeding the receptacle have a modified S-Curve when receptacle is in place.

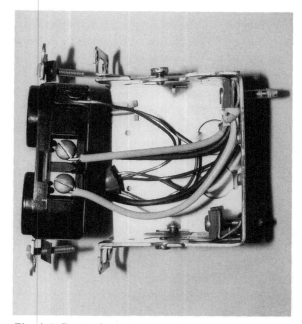

Fig. A-2. Receptacle ready to be pushed into box.

Fig. A-4. Receptacle mounted in box.

Fig. A-5. Auxiliary fuse panel. Requires 240V feeders, including a neutral. Time delay fuses (Fustats) are shown above the top of the panel. The adapter is in the center top.

fuses to be used in the regular fuse panel. The two arrows at left point to the two main terminals that feed the panel (240V). The feeder neutral goes all the way around to the right and under one of these screws on the neutral bar. These are all time delay fuses.

At the top of Fig. A-6 is a service-entrance cable box connector. The second row shows Romex connector, BX connector, Thinwall connector, and a Thinwall connector, compression type. The third row shows BX or greenfield connector, knockout plug, lock washer, and set-

cle mounting strap, there is no good ground. This is why the grounding wire goes to the receptacle grounding screw.

Figure A-4 shows the receptacle mounted in the box. Even though the mounting strap does not touch the box itself, the receptacle is effectively grounded. If the box was surface-mounted as a "handy box" is, then the receptacle mounting strap would contact the box itself and would not need a grounding wire to complete the ground.

EXPLANATION OF VARIOUS DEVICES

At the top of Fig. A-5 are type S fuses. In the center is the fuse adapter that allows these

Fig. A-6. Assortment of electrical fittings.

screw Thinwall connector. The fourth row shows Thinwall "one-hole" mounting clip, set-screw Thinwall connector, and Thinwall mounting clip. At the bottom is a piece of Romex *with* ground and a piece of Romex *without* ground. (Do not use Romex without a ground.)

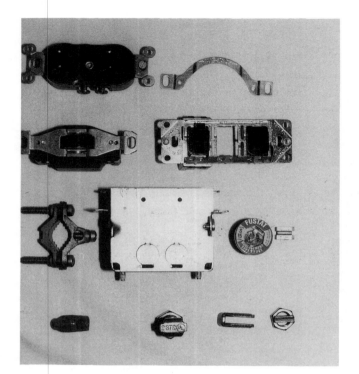

Fig. A-7. Assortment of wiring devices.

At the top of Fig. A-7 is a duplex receptacle, mounting strap for a plate such as for a phone, "snap" switch (this has a porcelain body—very rare), interchangeable devices (switch and receptacle), ground clamp for water pipe, box that is gangable, Fustat, pound on ground clamp for box, and the bottom row shows a wire nut, split clamp for heavy wires (notice the UL marking), and a clamp taken apart.

The following pages are excerpts from product listings and specifications offered by Leviton, followed by some junction boxes and their specifications, offered by Raco.

A3  LEVITON®

15 AMP

AC QUIET SWITCHES
120-277V AC, ½HP at 120V, 2 HP at 240V

53501-I

SPECIFICATION GRADE
*Heavy duty switches
are backed by a Limited
Ten-Year Warranty and
include these features:*

- Quiet, safe mechanical
action in any position.

- Large head terminal
screws backed out and
staked for fast wiring;
accept up to No. 10
copper or copper-clad
wire.

- Back-wiring clamps
accept up to No. 10
copper or copper-clad
wire.

- Large silver-cadmium
oxide contacts for
maximum conductivity.

- Heavy gauge rust
resistant steel mounting
strap.

- Shallow design for
maximum wiring room.

- Convenient washer type
break-off plaster ears for
best flush alignment.

- Captive mounting screws
for fast installation.

5501-I

5501-8I

5501-2I

BACK AND SIDE WIRED

Cat. No. Brown or Gray (-GY)	Cat. No. Ivory (-I) White (-W)	Description
53501	53501-I	Single pole, toggle
53502	53502-I	Double pole, toggle
53503	53503-I	3-Way, toggle
53504	53504-I	4-Way, toggle

SIDE WIRED

Cat. No. Brown or Gray (-GY)	Cat. No. Ivory (-I) White (-W)	Description
5501 5501-GY	5501-I	Single pole, toggle
5501-8	5501-8I	Single pole, Hospital Call*
5502 5502-GY	5502-I	Double pole, toggle
5502-8	5502-8I	Double pole, Hospital Call*
5503 5503-GY	5503-I	3-Way, toggle
5504 5504-GY	5504-I	4-Way, toggle

*Hospital Call Switches turn on when toggle is pulled down; have 3-foot braided cord

Cat. No. Brown or Gray (-GY)	Cat. No. Ivory (-I) White (-W)	Description	
5501-2 5501-2GY	5501-2I	Single pole, toggle	⎫
5502-2 5502-2GY	5502-2I	Double pole, toggle	⎬ GROUNDING
5503-2 5503-2GY	5503-2I	3-Way, toggle	
5504-2 5504-2GY	5504-2I	4-Way, toggle	⎭

All items above are UL
Listed, except where noted.

Illustrations on pages 241 through 279 are courtesy Leviton Co.

A4

15 AMP AC QUIET SWITCHES

SEE SECTION F FOR DECORA DESIGNER LINE SWITCHES

54501-W

SPECIFICATION GRADE
Heavy duty switches are backed by a Limited Ten-Year Warranty and include these features:

- Shallow design—less than 1 in. deep
- Framed toggle for faster alignment when ganging and dust protection
- Large silver-cadmium oxide contacts for maximum conductivity.

57501

SPECIFICATION GRADE

- Completely silent operation
- Red mounting screw at top for correct installation position.
- Shallow design for easy wiring
- Illuminated models have toggle that lights when switch is OFF
- Terminals accept up to No. 12 copper or copper clad wire.

5551-SP

All items above are UL Listed, except where noted.

SIDE WIRED
15A 120-277V AC ½ HP at 120V, 2 HP at 240V

Cat. No. Brown or Gray (-GY)	Cat. No. Ivory (-I) White (-W)	Description	
54501 54501-GY	54501-I 54501-W	Single pole	FRAMED TOGGLE
	54501-ISP	Single pole, SEE PACK	
54501-SP	54501-WSP		
54502 54502-GY	54502-I 54502-W	Double pole	
54503 54503-GY	54503-I	3-Way	
54503-SP	54503-ISP 54503-WSP	3-Way, SEE PACK	
54504 54504-GY	54504-I 54504-W	4-Way	
54504-2	54504-2I	4-Way with Grounding Screw	
	57501	Single pole, ivory	FRAMED TOGGLE, ILLUMINATED FRAME
	57503	3-Way, ivory	

SIDE WIRED
15A 120V AC ½ HP at 120V

5551-SP	5551-ISP 5551-WSP	Single Pole, toggle, SEE PACK	AC SILENT MERCURY
5553-SP	5553-ISP 5553-WSP	3-Way, toggle, SEE PACK	
5561-SP		Single pole, illuminated toggle SEE PACK	
5563-SP		3-Way, illuminated toggle SEE PACK	

A7

AC QUIET SWITCHES
120-277V AC, 1HP at 120V, 2 HP at 240V

SEE SECTION F FOR DECORA DESIGNER LINE SWITCHES

5521-I

54521

All items below are UL Listed, except where noted.

SPECIFICATION GRADE
Heavy duty switches are backed by a Limited Ten-Year Warranty and include these features:

- Quiet, safe mechanical action in any position.
- Large head terminal screws backed out and staked for fast wiring; accept up to No. 10 copper or copper-clad wire.
- Back-wiring clamps accept up to No. 10 copper or copper-clad wire.
- Large silver-cadmium oxide contacts for maximum conductivity.
- Heavy gauge rust resistant steel mounting strap.
- Shallow design for maximum wiring room.
- Convenient washer type break-off plaster ears for best flush alignment.
- Captive mounting screws for fast installation.

SIDE WIRED

Cat. No. Brown or Gray (-GY)	Cat. No. Ivory (-I) White (-W)	Description	
5521 5521-GY	5521-I	Single Pole, toggle	
5522 5522-GY	5522-I	Single Pole, toggle	RED COVER
5523 5523-GY	5523-I	3-Way, toggle	
5524 5524-GY	5524-I	4-Way, toggle	
5521-2 5521-2GY	5521-2I	Single Pole, toggle	
5522-2 5522-2GY	5522-2I	Double Pole, toggle	RED COVER GROUNDING
5523-2 5523-2GY	5523-2I	3-Way, toggle	
5524-2 5524-2GY	5524-2I	4-Way, toggle	
54521	54521-I 54521-W	Single Pole	
54522	54522-I 54522-W	Double Pole	RED COVER FRAMED TOGGLE
54523	54523-I 54523-W	3-Way	
54524	54524-I 54524-W	4-Way	

AC/DC SWITCHES

5302-I

SPECIFICATION GRADE
Heavy duty switches are backed by a Limited Ten-Year Warranty.

- All T-Rated switches control tungsten filament lamp loads up to full rated capacity.

FRONT WIRED
20A 125V T-RATED, 10A 250V

Cat. No. Brown	Cat. No. Ivory (-I)	Description
5301	5301-I	Single Pole, toggle
5302	5302-I	Double Pole, toggle
5303	5303-I	3-Way, toggle

B2 LEVITON

600 WATT
Decora
TOUCH DIMMER CONTROLS
120V 60Hz AC

6606

6607

SPECIFICATION GRADE

These General Duty Lighting Control units are backed by a Limited Two-Year Warranty and feature:

- Fits standard wall boxes, single or multi-gang
- Comes with matching Decora wallplate and may be ganged with any Decora devices
- 3-Way version offers complete dimming and ON/OFF at both stations
- Reliable solid-state circuitry with built-in radio/TV interference filter
- UL Listed

INCANDESCENT

Cat. No. Single Pole	Cat. No. 3-Way	Cat. No. 4-Way	Description
6606	6607*	6614	Brown Frame and Wallplate w/Gold Touch Plate
6606-I	6607-I*	6614-I	Ivory Frame and Wallplate w/Gold Touch Plate
6606-W	6607-W*	6614-W	White Frame and Wallplate w/Silver Touch Plate

*Includes both units for 3-Way circuit. Follow wiring diagrams EXACTLY. Not for use in 4-way circuits.

NOTE: When two units are ganged together, each unit should not be loaded beyond 500 watts. When three or more units are ganged, each unit should not be loaded beyond 400 watts.

FLICKERING MAY OCCUR IN three phase power systems. Connect all dimmers to the same phase or run separate neutral to each phase.

6608

6609

Sensitron
TOUCH DIMMER CONTROLS

SPECIFICATION GRADE

These General Duty Lighting Controls are backed by a Limited Two-Year Warranty and feature:

- Built-in electronic memory retains light level setting until readjustment
- Fits standard wall box
- Soft illumination at base of plate on single-pole models makes location in darkness easy
- Saves energy, extends bulb life
- Built-in radio/TV interference filter
- Unique design compliments any interior
- UL Listed

FULL RANGE CONTROL WITH NO MOVING PARTS

Just touch to operate   Touch once—it's on
Touch and hold—for full range dimming   Touch again—it's off

*3-Way SENSITRON offers complete dimming AND on/off control at BOTH stations. This PAIR of units must be used together—not with any other 3-Way switches and/or dimmers.

INCANDESCENT
600W 120V AC

Cat. No. Single Pole	Cat. No. 3-Way	Description
6608	6609*	Brown with Gold Touch Plate and Trim
6608-I	6609-I*	Ivory with Gold Touch Plate and Trim
6608-W	6609-W*	White with Silver Touch Plate and Trim
6608-E	6609-E*	Black with Silver Touch Plate and Trim

NOTE: SENSITRON may be ganged 2 units per 3-gang box ONLY. When doing so, each unit must be derated to 500 Watts maximum.

*Includes both units for 3-Way circuit. Follow wiring diagrams EXACTLY. Not for use in 4-way circuits.

FLICKERING MAY OCCUR IN three phase power systems. Connect all dimmers to the same phase or run separate neutral to each phase.

600 WATT
Decora
SLIDE DIMMERS
AND MOTOR SPEED CONTROL

LEVITON **B3**

6621   6627

STANDARD GRADE
These general duty devices are backed by a Limited Two-Year Warranty and feature:

- Positive ON/OFF switching contacts with upper end dimming bypass for maximum brightness (Incandescent Only)
- Large, durable silver cadmium oxide contacts
- Compact design for maximum space in wall box
- Solid-state circuitry with built-in radio/TV interference filter
- Permanently mounted slide switch
- Can be ganged with other units †
- Illuminated version available (white and ivory only)
- UL Listed (Dimmers Only) Speed Controls UL Classified

Cat. No.	Description	Rating	Colors
6621	Standard Slide Dimmer	600W 120V 5A 60 Hz AC Only	Brown, Ivory (-I), White (-W)
6631	Illuminated Slide Dimmer	600W 120V 5A 60 Hz AC Only	Ivory (-I), White (-W)
6627-1	Slide Fan Speed Control w/Manual Low Speed Trim Adjustment	Single pole 5A 120V 60 Hz AC Only	Brown, Ivory (-I), White (-W)

†When two units are ganged together, each unit should not be loaded beyond 500 watts. When three or more units are ganged, each should not be loaded beyond 400 watts.

600 WATT
Motor
SPEED CONTROLS

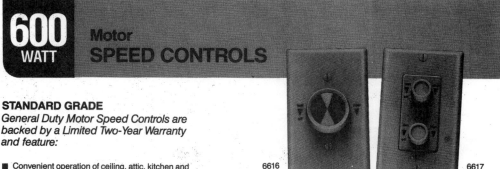

6616   6617

STANDARD GRADE
General Duty Motor Speed Controls are backed by a Limited Two-Year Warranty and feature:

- Convenient operation of ceiling, attic, kitchen and bathroom fan units with split capacitor or shaded pole motors ONLY
- Low speed trim adjuster insures proper restart after power interruption to prevent motor overheating (Cat. No. 6616 Only.)
- Reduced fan noise at low settings
- Positive ON/OFF switching, turning fans on at high speed
- Fits standard wall boxes; can be ganged with other devices**
- Furnished with indicated ivory wallplate, mounting screws and stripped, tinned 4¾ in. 18 AWG wire leads for fast installation
- UL Classified for Specific Fan Manufacturers

Cat. No.	Rating	Description
6616-1XI (Ivory) 6616-1XW (White)	600W 5A 120V AC 60 Hz Only	Wall Mounted Motor Speed Control—operates 1 or 2 units within rating limits
6617-XI (Ivory) 6617-XW (White)	Speed Control— 300W 2.5A Dimmer— 300W 2.5A 120V AC 60Hz Only	Wall Mounted Motor Speed Control w/ Full Range Dimmer operates one fan and one light within rating limits

**These devices are not replacements for standard switches. Additional wiring may be necessary for proper fan operation, depending on the manufacturer's individual specifications.

245

B4 LEVITON

600 WATT

TRIMATRON DIMMERS
600W 120V AC

trimatron®

Conventional dimmer

6681

6602

6603

STANDARD GRADE
These General Duty Lighting Control units are backed by a Limited Two-Year Warranty and feature:

- Solid state electronics assure reliability
- Built-in radio/TV interference filter
- Wall mounting types fit single gang outlet boxes
- Ultra-compact—only half as deep as conventional dimmers
- Complete with leads for fast installation

DELUXE PUSH-ON/PUSH-OFF

Cat. No.	Description
6681	Single Pole
6681-2	Single Pole w/ Grounding Lead
800-6681-BP	Single Pole—SLIMPAK Carded
6683	3-Way
6683-2	3-Way w/ Grounding Lead
800-6683-BP	3-Way—SLIMPAK Carded

ROTARY

Cat. No.	Description
6602	Single Pole
6602-2	Single Pole w/ Grounding Lead
800-6602-BP	Single Pole, SLIMPAK Carded

ECONOMY DIMMERS
FULL RANGE, PUSH-ON/PUSH OFF

Cat. No.	Description
6600	Single Pole
6600-BP	Single Pole, Blister-Pak Carded

NOTE: When two units are ganged together, each unit should not control more than 500 watts. When three or more units are together, each unit should not control more than 400 watts.

Low WATT

TRIMATRON HI-LO
300W 120V AC

STANDARD GRADE
These General Duty Lighting Control units are backed by a Limited Two-Year Warranty and feature:

- Three positions—OFF, HI (full brightness), LO (33% brightness)

HI-LO

Cat. No.	Description
6603	Single Pole
800-6603-BP	Single Pole, SLIMPAK Carded

26115-W

DIMMER KNOBS
with metal insert
Packed 25 per box

26115 ˙ – Brown
26115-I – Ivory
26115-E – Ebony
26115-W – White

All items above are UL Listed, except where noted.

15 AMP

RECEPTACLES
125V, 2-POLE, 3-WIRE; 2-WIRE

LEVITON C5

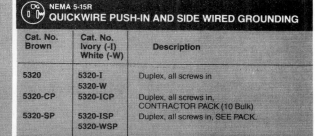

5320-I

5248-X

1215-I

1228

NOTE: SEE PACK is a single unit box with a transparent window for product visibility.

CONTRACTOR PACK is a box of 10 bulk-packed devices (no individual packaging) for speedy on-the-job installation.

STANDARD GRADE
General duty receptacles are backed by a Limited Two-Year Warranty.

All items above are UL Listed, except where noted.

NEMA 5-15R — QUICKWIRE PUSH-IN AND SIDE WIRED GROUNDING

Cat. No. Brown	Cat. No. Ivory (-I) White (-W)	Description
5320	5320-I 5320-W	Duplex, all screws in
5320-CP	5320-ICP	Duplex, all screws in, CONTRACTOR PACK (10 Bulk)
5320-SP	5320-ISP 5320-WSP	Duplex, all screws in, SEE PACK.

NEMA 5-15R — QUICKWIRE PUSH-IN AND SIDE WIRED

Cat. No. Brown	Cat. No. Ivory (-I) White (-W)	Description
5320-S	5320-SI	Duplex, Self-Grounding
	5320-SW	
5320-5	5320-5I	Duplex, one screw backed out on each side CONTRACTOR PACK (10 Bulk)
5320-6	5320-6I	Duplex, all screws backed out CONTRACTOR PACK (10 Bulk)

NEMA 5-15R — 8-HOLE QUICKWIRE PUSH-IN GROUNDING

Cat. No. Brown	Cat. No. Ivory (-I) White (-W)	Description	
5248-CPX	5248-ICP	Duplex, (10 Bulk) CONTRACTOR PACK	ACCEPTS UP TO NO. 10 WIRE
5248-4X	5248-4IX	Duplex, less plaster ears	

NEMA 5-15R — SIDE WIRED

Cat. No. Brown	Cat. No. Ivory (-I) White (-W)	Description
1215	1215-I 1215-W	Duplex
1215-CP	1215-ICP	Duplex, CONTRACTOR PACK (10 Bulk)
1215-SP	1215-ISP 1215-WSP	Duplex, SEE PACK (Single unit box w/ plastic window for product visibility)
1215-4	1215-4I	Duplex, less plastic ears
1228		Duplex, on 4 in. cover

C6 LEVITON

15 AMP

RECEPTACLES

125V, 250V 2-POLE, 3-WIRE GROUNDING

2650-I

UL LISTED CO/ALR

CO-ALR STANDARD GRADE
• *UL Listed*
• *For direct connection to No. 12 or No. 10 aluminum conductors*

All items below are UL Listed, except where noted.

STANDARD GRADE
General duty receptacles are backed by a Limited Two-Year Warranty.

NOTE: SEE PACK is a single unit box with a transparent window for product visibility.

223

5000

STANDARD GRADE
■ Single receptacle on recessed wallplate with strong hook for hanging heavy clocks.

658

Spec-Master ®
PREMIUM
SPECIFICATION GRADE

5024

NEMA 5-15R CO/ALR SIDE WIRED

Cat. No. Brown	Cat. No. Ivory (-I) White (-W)	Description
2650	2650-I	Duplex
	2650-W	
2650-S	2650-SI	Duplex, Self-Grounding

NEMA 1-15R SIDE WIRED NON-GROUNDING

238	238-I	Duplex, one screw each side backed out	
223	223-I	Duplex	2 WIRE
223-SP	223-ISP	Duplex, SEE PACK	
	223-WSP		
222	222-I	Duplex, less plaster ears	
229		Duplex, on 4 in. cover	
242	242-I	Duplex	T-SLOT— Not UL Listed For replacement only
5000	5000-I	Duplex	
5229		Duplex, on 4 in. cover	
628	628-I	Clockhanger, one-piece smooth design	2-POLE 2-WIRE

NEMA 5-15R SIDE WIRED, CLOCKHANGER

658	658-I	Clockhanger, aluminum finish, smooth wallplate
658-BR		Clockhanger, brass finish, smooth wallplate
688	688-I	Clockhanger, one-piece, smooth design

3-WIRE POLARIZED SIDE WIRED

Cat. No. Black	Cat. No. Ivory (-I)	Description
5024	5024-I	Single (Not UL Listed. For replacement use only)

NEMA 6-15R 250V SIDE WIRED

5651	5651-I	Single, meets latest Fed. Spec. W-C-596F

20 AMP

RECEPTACLES

125V, 2-POLE, 3-WIRE

5800-I

5801-I

2660-I

5842-I

5032-I

All items below are UL Listed, except where noted.

SPECIFICATION GRADE
Heavy duty receptacles are backed by a Limited Ten-Year Warranty.

CO-ALR STANDARD GRADE
See page C6 for features

SPECIFICATION GRADE
Heavy duty receptacles are backed by a Limited Ten-Year Warranty.

NEMA 5-20R BACK AND SIDE WIRED GROUNDING

Cat. No. Brown or Gray (-GY)	Cat. No. Ivory (-I) or White (-W)	Description
5896	5896-I	Duplex
5891	5891-I	Single

NEMA 5-20R SIDE WIRED GROUNDING

5800	5800-I	Duplex
5800-GY		
5800-SP	5800-ISP	Duplex, SEE PACK (single unit box w/ plastic window for product visibility)
5803		Duplex, on 4 in. cover
5800-4	5800-4I	Duplex, less plaster ears
5800-S	5800-SI	Duplex, Self-Grounding
5801	5801-I	Single
	5801-W	
5805		Single, on 4 in. cover

NEMA 5-20R CO/ALR SIDE WIRED GROUNDING

Cat. No. Brown	Cat. No. Ivory (-I)	Description
2660	2660-I	Duplex

SPECIAL DUAL VOLTAGE 125/250V

NEMA 5-20R BACK AND SIDE WIRED — NEMA 6-20R GROUNDING

5844	5844-I	Duplex, dual voltage

SIDE WIRED GROUNDING

5842	5842-I	Duplex, dual voltage

NEMA 10-20R SIDE WIRED NON-GROUNDING

Cat. No. Black	Cat. No. Ivory (-I)	Description
5032	5032-I	Single
5034		Single, on 4 in. cover

 C10 **LEVITON**

20 AMP

RECEPTACLES
250V, 2-POLE, 3-WIRE, GROUNDING

5461

Spec-Master®
**PREMIUM
SPECIFICATION GRADE**
*Top-of-the-line
receptacles are backed
by a Limited Ten-Year
Warranty.*

(See page C3 for details)

5824

SPECIFICATION GRADE
*Heavy duty receptacles
are backed by a Limited
Ten-Year Warranty and
feature:*

- Large-head terminal
  screws backed out and
  staked, accept up to No. 10
  copper or copper-clad wire.

- Quickwire™ push-in
  terminals provide fastest
  wiring; accept up to No. 12
  copper or copper-clad wire.

- All power contacts
  are double wipe for
  maximum conductivity
  and plug retention.

- Heavy gauge, rust-resis-
  tant steel mounting strap.

- Shallow design for maximum wiring room in box.

- Convenient washer-type
  break-off plaster ears for
  best flush alignment.

- Captive mounting screws
  for fast installation.

- Back-wiring clamps
  accept up to No. 10 copper
  or copper-clad wire.

- Break-off fins on duplex
  receptacles allow easy
  two-circuit conversion.

5821-I

NEMA 6-20R   ANSI C73.51
BACK AND SIDE WIRED

Cat. No.	Description
5462	Duplex
5461	Single

NOTE: Catalog numbers above are Brown. For other colors, add
suffix letters to basic cat. no. as follows: -I (Ivory), -W(White),
-R(Red), -G(Gray), -Y (Yellow—duplex only).

NEMA 6-20R
BACK AND SIDE WIRED

Cat. No. Brown	Cat. No. Ivory (I) White (-W)	Description
5824	5824-I	Duplex
5823	5823-I	Single

SIDE WIRED

5822	5822-I	Duplex
5827		Duplex, on 4 in. cover
5821	5821-I	Single
	5821-W	
5821-SP	5821-ISP	Single, SEE PACK (single unit box w/plastic window for product visibility)
5825		Single, on 3¼ in. cover
5826		Single, on 4 in. cover

All items above are UL
Listed, except where noted.

30 AMP

POWER RECEPTACLES
125V, 250V, 125V/250V

LEVITON **C11**

5371

5375

5054

55054

SPECIFICATION GRADE
Heavy duty receptacles are backed by a Limited Ten year Warranty and feature:

- UL Listed for copper or aluminum conductors (clearly marked "AL/CU").

- Heavy gauge double wipe bronze contacts for maximum conductivity.

- Color coded terminals for easy identification and fast wiring.

- All mounting hardware included

- Flush mount devices fit single or two-gang outlet boxes (except cat. nos. 278, 279, 5051, 5055).

- Surface mount devices have concentric knockouts for 3/4 in. and 1 in. conduit and adjustable cord clamp and back and bottom entrance ports for conductor cable.

NEMA 5-30R 125V
2-POLE, 3 WIRE, GROUNDING

Cat. No. Black	Cat. No. Gray (-GY) White (-W)	Description	Terminal Type	Wire Gauge (AWG)
5371		Flush mount Receptacle	Pressure	4 max.
5375	5375-GY*	Surface Mount Receptacle	Lay-In	4 max.
7313		Flush Mount Receptacle for recreational vehicles -ANSI Std. C73.13	Pressure	4 max.

NEMA 6-30R 250V
2-POLE, 3-WIRE, GROUNDING

5372		Flush Mount Receptacle	Pressure	4 max.
5376	5376-GY*	Surface Mount Receptacle	Lay-In	4 max.

NEMA 10-30R 125/250V
3-POLE, 3-WIRE, NON-GROUNDING

5207	5207-GY*	Flush Mount Receptacle	Pressure	4 max.
5055		Flush Mount Receptacle for 4 in. square outlet	Lay-In	4 max.
5054-2	5054-2W*	Surface Mount Receptacle	Lay-In	4 max.
5054		Surface Mount Receptacle	Screw	4 max.

NEMA 14-30R 125V/250V
3-POLE, 4-WIRE, GROUNDING

278	278-GY*	Flush Mount Receptacle	Pressure	4 max.
55054	55054-W*	Surface Mount Receptacle	Lay-In	4 max.
276		Panel Mount Receptacle	Pressure	4 max.

*When existing stock is depleted, these items will be DISCONTINUED

See page E5 for Power Receptacle Wallplates

All items above are UL Listed, except where noted.

C12 LEVITON

50 AMP

POWER RECEPTACLES
125V, 250V, 125V/250V

5378

5206

5051

55050

SPECIFICATION GRADE
Heavy duty receptacles are backed by a Limited Ten year Warranty and feature:

- UL Listed for copper or aluminum conductors (clearly marked "AL/CU").

- Heavy gauge double wipe bronze contacts for maximum conductivity.

- Color coded terminals for easy identification and fast wiring.

- All mounting hardware included

- Flush mount devices fit single or two-gang outlet boxes (except cat. nos. 278, 279, 5051, 5055).

- Surface mount devices have concentric knockouts for 3/4 in. and 1 in. conduit and adjustable cord clamp and back and bottom entrance ports for conductor cable.

NEMA 5-50R 125V
2-POLE, 3-WIRE, GROUNDING

Cat. No. Black	Cat. No. Gray(-GY) White (-W)	Description	Terminal Type	Wire Gauge (AWG
5373	5373-GY*	Flush Mount Receptacle	Pressure	4 ma
5377	5377-GY*	Surface Mount Receptacle	Lay-In	4 ma

NEMA 6-50R 250V
2-POLE, 3-WIRE, GROUNDING

5374	5374-GY*	Flush Mount Receptacle	Pressure	4 max
5378	5378-GY*	Surface Mount Receptacle	Pressure	4 max

NEMA 10-50R 125V/250V
3-POLE, 3-WIRE, NON-GROUNDING

5206	5206-GY*	Flush Mount Receptacle	Pressure	4 max.
5051		Flush Mount Receptacle for 4 in. square outlet box	Lay-In	4 max.
5050	5050-W*	Surface Mount Receptacle	Lay-In	4 max.

NEMA 14-50R 125V/250V
3-POLE, 4-WIRE, GROUNDING

279	279-GY*	Flush Mount Receptacle	Pressure	4 max.
55050	55050-W*	Surface Mount Receptacle	Lay-In	4 max.
277		Panel Mount Receptacle wired ground side	Pressure	4 max.
277-7		Panel Mount Receptacle wired neutral side	Pressure	4 max.

*When existing stock is depleted, these items will be DISCONTINUED

See page E5 for Power Receptacle Wallplates

All items above are UL Listed, except where noted.

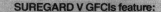

15 AMP

GROUND FAULT CIRCUIT INTERRUPTERS

LEVITON **D1**

All Specification Grade and Standard Grade GFCIs feature:

- Conformation with UL Standard 943 Class A
- Complete testing of each device before shipping
- Tough, impact resistant construction
- Temperature tolerance level of −31°F to 158°F
- Feed through ready
- Matching wallplate included (except where noted)
- Compatibility with Decora wallplates

6598-I

STANDARD GRADE
General Duty GFCIs are backed by a Limited Two-Year Warranty

SUREGARD V GFCIs feature:

- 1⅛ in. deep body
- Rounded corners
- Vertical plug slots
- Mounting strap with break-off plaster ears
- Terminal screws with silver alloy terminal contacts
- Trip threshold of 5mA.
- Trip time of 0.025 seconds

SUREGARD V   NEMA 5-15R
15A 125V at receptacle; 20A 125V feed-through

Cat. No. Brown	Cat. No. Ivory	Cat. No. White	Description
6598	6598-I	6598-W	With Indicator Light
6599	6599-I	6599-W	No Indicator Light
6599-X	6599-XI		No Indicator Light, No Wallplate, CONTRACTOR PACK
6599-L	6599-LI		With Leads

HOSPITAL GRADE GFCIs feature:

- UL LISTED Hospital Grade High Abuse Receptacle construction
- Certified Corrosion Resistant with cupro-nickel exposed metal parts
- 1⅛ in. deep body
- Horizontal plug slots
- Pre-stripped wire leads for fast installation

HOSPITAL GRADE   NEMA 5-15R
15A 125V at receptacle

Cat. No. White	Cat. No. Ivory	Description
6198-HGW	6198-HGI	LED Indicator Light, Nylon Face

6198-HGI

SPECIFICATION GRADE
These heavy duty GFCIs are backed by a Limited Two-Year Warranty

F9

HOME CONTROL SYSTEM

Catalog Number	Description
6321	**Wall-Mounted Programmer**—Automatically carries out programmed scheduling for 8 addresses. Two distinct on/off cycles can be programmed for each address in a 24-hour period. Full manual override capability for all 8 addresses will not affect programmed memory. Memory is protected by a 15 hr. battery back-up to preserve programmed scheduling and clock time in the event of power interruption (batteries not included). Surface mounts to a standard single-gang wall box using two 152.4mm (6") leads for easy installation. Rated 120V 60 Hz AC only. Ivory only.
6320	**Table Top Controller**—provides manual control of all 16 Number Codes with Letter Code selection dial for A-P on back. Console under hinged lid has buttons for transmitting ON, OFF, DIM, BRIGHTEN, ALL LIGHTS ON and ALL LIGHTS OFF commands. Plugs into any 125V AC outlet. Brown only.
6319-2D, -2DW **6319-4** **6319-4D, -4DW** **6319-4A**	**Wall-Mounted Controller**—Wall-mounted controllers are used to exercise manual remote control and are available in four versions: **6319-2D**—Provides on/off control and dimming for one address. **6319-4**—Provides manual on/off control for four addresses. **6319-4D**—Provides manual on/off control and dimming for three addresses. **6319-4A**—Provides on/off control for three addresses and also provides ALL LIGHTS ON and ALL OFF command capability for all designated addresses in the Home Control System Network. Up to four wall-mounted controllers can be ganged to control sixteen addresses. Wall-mounted controllers are packed with a single-gang matching decorator wall-plate. They mount in a standard single-gang wall box using 101.6mm (4") leads for easy installation. Rated 120V 60 Hz AC only. Ivory and White only.
6326	**Interflash Controller**—automatically sends intermittent ON/OFF signals to all system Switch Modules and Receptacle Modules set to the same Letter and Number Code as the Interflash. Responds to an input signal of 6 to 24 volts DC. When the DC input terminates, all lights will remain ON until turned OFF by a system controller or manually at individual modules. Ivory only. Rating—Line Side: 120V AC 60 Hz Only Control Side: Up to 24V DC Only

All components are UL Listed and backed by Leviton's limited one year warranty.

5211-I

5218-I

5214-I

5227-I

5213-LBI

5 10 15 AMP

COMBINATION and INTERCHANGEABLE DEVICES

LEVITON G1

SEE SECTION F FOR DESIGNER LINE COMBINATION DEVICES

AC/DC T-RATED AND AC COMBINATION DEVICES

- Available with AC Quiet Switches or AC/DC Switches
- Devices with AC Quiet Switches have break-off fins for conversion to separate feeds (except switch/pilot light combinations)
- AC Quiet Switches have silver cadmium oxide contacts and receptacles have double-wipe contacts for maximum conductivity
- All devices side wired with terminal screws that accept up to No. 10 copper or copper clad wire
- All devices fit standard duplex wallplates
- All switches have framed toggles to reduce dust accumulation

All items below are UL Listed, except where noted.

SPECIFICATION GRADE
These heavy duty devices are backed by a Limited Ten-Year Warranty.

+ 250V Switch rating not available

NOTE: SEE PACK is a single unit box with a transparent window for product visibility.

Pilot lights are not covered by warranty.

NEMA 1-15R RECEPTACLE 15A 125V
SWITCH 10A 125V T, 5A 250V

Cat. No. Brown	Cat. No. Ivory (-I)	Description	Feed
5211 +	5211-I	Single Pole Switch and Receptacle	Common
5211-SP	5211-ISP	Same as above, SEE PACK	
5221	5221-I	Single Pole Switch and Receptacle	Separate

NEMA 5-15R RECEPTACLE 15A 125V
SWITCH 10A 125V T, 5A 250V

5219	5219-I +	Single Pole Switch and Grounding Receptacle	Common
5220	5220-I	Single Pole Switch and Grounding Receptacle	Separate

SWITCHES 10A 125V T, 5A 250V

5212	5212-I	2 Single Pole Switches	Common
5212-SP	5212-ISP	Same as above, SEE PACK	
5212-2	5212-2I	2 Single Pole Switches, Grounding	Common
5214	5214-I	2 Single Pole Switches	Separate
5215	5215-I	1 Single Pole, 1 3-Way Switch	Common
5216	5216-I	1 Single Pole, 1 3-Way Switch	Separate
5217	5217-I	2 3-Way Switches	Common
5218	5218-I	2 3-Way Switches	Separate

SWITCH 10A 125V T **NEON PILOT** 1/25W 125V

5227	5227-I	Single Pole Switch and Neon Pilot Light	Common

SWITCH 10A 125V T
INCANDESCENT PILOT 75W 125V

5213-LB	5213-LBI	Single Pole Switch and Incandescent Pilot Light, Less Bulb	Common

G2 LEVITON

15 20 AMP

COMBINATION and INTERCHANGEABLE DEVICES
SEE SECTION F FOR DESIGNER LINE COMBINATION DEVICES

5222-I

5225-I

5224-I

5226-I

SPECIFICATION GRADE
These heavy duty devices are backed by a Limited Ten-Year Warranty.

All items above are UL Listed, except where noted.

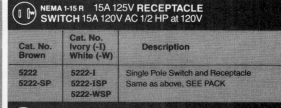

NEMA 1-15 R 15A 125V **RECEPTACLE**
SWITCH 15A 120V AC 1/2 HP at 120V

Cat. No. Brown	Cat. No. Ivory (-I) White (-W)	Description
5222	5222-I	Single Pole Switch and Receptacle
5222-SP	5222-ISP	Same as above, SEE PACK
	5222-WSP	

 NEMA 5-15 R 15A 125V **RECEPTACLE**
SWITCH 15A 120V AC 1/2 HP at 120V

5225	5225-I	Single Pole Switch and Grounding Receptacle
5225-SP	5225-ISP	Same as above, SEE PACK
	5225-WSP	
5245	5245-I	3-Way Switch and Grounding Receptacle
	5245-W	

SWITCH 15A 120-277V AC, 1/2 HP at 120V, 2 HP at 240V

5224	5224-I	Two Single Pole Switches
5224-SP	5224-ISP	Same as above, SEE PACK
	5224-WSP	
5224-2	5224-2I	Two Single Pole Switches, Grounding
	5224-2W	
5241	5241-I	Single Pole and 3-Way Switch
	5241-W	
5243	5243-I	Two 3-Way Switches
	5243-W	

SWITCH 15A 120V AC 1/2 HP at 120V
NEON PILOT LIGHT 1/25W-125V

5226	5226-I	Single Pole Switch and Neon Pilot Light
	5226-W	
5226-SP	5226-ISP	Same as above, SEE Pack
	5226-WSP	

SWITCH 20A 120/277V AC 1/2 HP at 120V 2 HP at 240V

5334	5334-I	Two Single Pole Switches

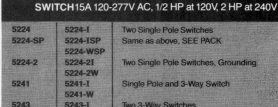

NEMA 5-20 R 20A 120V **RECEPTACLE**
SWITCH 20A 120V

5335	5335-I	Single Pole Switch and Grounding Receptacle

SWITCH 20A 120V AC 1/2 HP at 120V
NEON PILOT 1/25W 125V

5336	5336-I	Single Pole Switch and Neon Pilot Light

5 10 15 AMP

COMBINATION and INTERCHANGEABLE DEVICES

SEE SECTION F FOR DESIGNER LINE COMBINATION DEVICES

LEVITON G3

INTERCHANGEABLE DEVICES

See page E-5 for complete listing of interchangeable device wallplates.

- Heavy duty phenolic or urea housings; switches protected from dust
- Completely interchangeable with other lines of similar devices conforming to NEMA specifications
- Devices lock onto steel mounting straps without mounting screws
- Terminal screws and Quickwire™ Push-In Terminals accept up to No. 12 copper or copper clad wire
- Switches and mounting straps meet Federal Specification W-S-893c

1790-I

AC QUIET SWITCHES

QUICKWIRE™ PUSH-IN AND SIDE WIRED
15A 120-277V AC 1/2 HP at 120V 2HP at 240V

STANDARD GRADE
These general duty devices are backed by a Limited Two Year Warranty

See page E5 for a complete selection of straight line plastic and satin finish metal wallplates for interchangeable devices.

Additional interchangeable devices listed on page L5.

Cat. No. Brown	Cat. No. Ivory (-I)	Description
1760	1760-I	Single Pole Switch, Quickwire
1761	1761-I	3-Way Switch, Quickwire
1790*	1790-I*	Single Pole Switch, Side Wired
1791*	1791-I*	3-Way Switch, Side Wired

AC/DC T-RATED SWITCHES
SIDE WIRED 5A 125V T 2A 250V

1752-T	1752-TI	4-Way Switch

10A 125V T 5A 250V

1750-T	1750-TI	Single Pole Switch
1751-Y	1751-TI	3-Way Switch

AC/DC SWITCHES
SIDE WIRED 10A 125V 5A 250V

1750	1750-I	Single Pole Switch
1756	1756-I	Double Pole Switch
1751	1751-I	3-Way Switch

NEMA 5-15R **SIDE WIRED** 15A 125V
2-POLE, 3-WIRE GROUNDING RECEPTACLE

1787	1787-I	Single

1760-I

1751-T

1787

All items above are UL Listed, except where noted.

*When existing stock is depleted, this item will be DISCONTINUED.

2-Pole 2-Wire | STRAIGHT BLADE PLUGS CONNECTORS AND INLETS

H1

All devices in this section are DEAD FRONT or "...so constructed that there are no exposed current carrying parts except the prongs, blades or pins" (1978 National Electrical Code, Article 410-56 (d), as adopted by Underwriters Laboratories, Inc.).

- Plug blades and connector contacts are brass for maximum conductivity.
- Armored devices have heavily plated steel shields and cord clamps.
- Vinyl devices resist chipping, cracking, oils, grease and acids.
- Rubber devices resist cold, abrasion, grease, oils, acids and impact.
- Plastic devices resist grease, oils and acids.

123

STANDARD GRADE
These general duty devices are backed by a Limited Two-Year Warranty.

- *Packed 500 Bulk

111   101

- See display page H3 for Cat. Nos. 105 and 107
- *Packed 500 Bulk

102   48643

48646   112

48642

+ Not UL Listed. For replacement use only.

All items below are UL Listed, except where noted.

| 10A 125V w/ No. 18-2 SPT-1 3A 125V w/ No. 20-2XT | **PARALLEL CORDS ONLY** |

NEMA 1-15P Non-polarized   NEMA 1-15P Polarized

Cat. No.	Description	Color	Cord Dia. (inches)	Wire Gauge (stranded)
123	Flat Plastic Easy-Wire Plug—Packed 2 per card	Brown		
123-I	Same as No. 123	Ivory	No. 18-2 SPT-1 or No. 20-2 XT Only	
123-P*	Same as No. 123, POLARIZED	Brown		
123-PI*	Same as No. 123-P	Ivory		

NEMA 1-15P 15A 125V   NEMA 1-15R   (Indicated below by +)

Cat. No.	Description	Color	Cord Dia. (inches)	Wire Gauge (stranded)
101	Clamptite Plug	Brown White	No. 20, 18, 16 Flat	
102	Clamptite Connector	Black	No. 18-SV or -SVT Round	
111	Rubber Plug, w/ vinyl blade, terminal, cord clamp inner assembly	Black	.296-.562	18.14
112	Rubber Connector mates w/ No. 111 above	Black	.296-.562	18-14
+5087	Rubber Connector, cord clamp	Black	.296-.562	14 max.
48643	Short Flat Vinyl Plug	Brown White (-W) Black (-E)	.375 max.	18-14
+638*	Angle Vinyl Plug	Black Brown (-B) White (-W)	.343 max.	16 max.
+1511*	Oval Vinyl Handle Connector	Black	.375 max.	14 max.
48646	Round Vinyl Handle Plug, cord clamp	Black	.437 max.	18-14
+612	No. 610 Connector, cord clamp	Black	.437 max.	14 max.
48642	Round Vinyl Handle Plug, spring blades	Black	.375 max.	18-14
1507*	Rubber Handle Plug, Continental pin blades	Black	.406 max.	14 max.

OUTLET BOX LAMPHOLDERS

 J1

8827-C

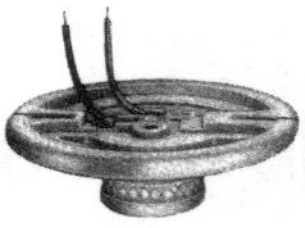

9875-2

9883

9870

Captive Mounting Screws

29816-C

All items below are UL Listed, except where noted.

STANDARD GRADE
These general duty devices are backed by a Limited Two-Year Warranty.

■ Medium bases

■ Shadeholder grooves

■ 4-Terminal models to facilitate continuous wiring

■ Pull chain models have bell at end of chain or cord

STANDARD GRADE
■ Removable Interior Mechanism w/Captive Mounting Screws

ONE-PIECE KEYLESS TOP WIRED
600W 600V

Cat. No. Brown Phenolic	Cat. No. White Urea	Terminals	O.D.	Fits Box Size
8829-C	8829-W	4 Screws	4½ in.	3¼ in., 4 in.

TWO-PIECE PULL CHAIN TOP WIRED
250W 250V (removable interior mechanism)

8827-C	8827-W	4 Screws	4½ in.	3¼ in., 4 in.

ONE-PIECE KEYLESS TOP WIRED
660W 250V

Cat. No. Porcelain	Terminals	O.D.	Fits Box Size
9874	2 Screws	3¾ in.	3¼ in.
9875	2 Screws	4½ in.	3¼ in., 4 in.
9975	2 Screws	5¼ in.	3¼ in., 4 in.
9975-2	6 in. pigtail leads	5¼ in.	3¼ in., 4 in.
9875-2	6 in. pigtail leads	4½ in.	3¼ in., 4 in.
49875	4 Screws	4½ in.	3¼ in., 4 in.
9883	4 Quickwire push-in	4½ in.	3¼ in., 4 in.

COVER MOUNTED KEYLESS TOP WIRED
660W 600V

9870	2 Screws	3½ in.	3¼ in.
9931	2 Screws	4⅛ in.	4 in.
9871	6 in. pigtail leads	3½ in.	3¼ in.

660W 250V

9873	6 in. pigtail leads	4⅛ in.	4 in.

TWO-PIECE PULL CHAIN TOP WIRED
250W 250V

29816-C	2 Screws	4½ in.	3¼ in., 4 in.
29816-CM	Interior mechanism for #s 29816, 29916, 49816		
29916-C	2 Screws	5¼ in.	3¼ in., 4 in.
29916-C2	6 in. pigtail leads	5¼ in.	3¼ in., 4 in.
49816-C	4 Screws	4½ in.	3¼ in., 4 in.

J2

15 AMP

OUTLET BOX LAMPHOLDERS

9816-C

9716-C

STANDARD GRADE
These general duty devices are backed by a Limited Two-Year Warranty.

- With 2-Pole 2-Wire Outlet
- Removable Interior Mechanism

- Used for pull-chain lampholders and for hanging fixture bowls.

TWO-PIECE PULL CHAIN SIDE WIRED
250W 250V

Cat. No. Porcelain	Terminals	O.D.	Fits Box Size
9816-C	2 Screws	4½ in.	3¼ in., 4 in.
9816-C2	6 in. pigtail leads	4½ in.	3¼ in., 4 in.

NEMA 1-15R LAMPHOLDER 250W 125V OUTLET 15A 125V
TWO-PIECE PULL CHAIN SIDE WIRED

9715-C	2 Screws	3¾ in.	3¼ in.
9716-C	2 Screws	4½ in.	3¼ in., 4 in.

NEMA 5-15R LAMPHOLDER 660W 125V OUTLET 15A 125V
TWO-PIECE PULL CHAIN TOP WIRED

9726	2 Screws	4½ in.	3¼ in., 4 in.
9726-CM	Interior Mechanism Only		

9726

PULL CHAIN OUTLET BOX LAMPHOLDERS
Suffix letter denotes type of pull chain attachment as illustrated.
(Type C listed above— for other types, substitute letter suffix
-A or -B as desired.)

Type A:
Chain with bell 5 inches long.

Type B:
Short chain and insulator with 5 inches of chain and bell attached.

Type C:
Short chain with connector and 3 feet of braided cord and bell.

8302-1

 8373   9820

6

8376   8375

9839

All items above are UL Listed, except where noted.

BALL CHAIN AND ACCESSORIES

Cat. No. Brass Finish	Cat. No. Nickel Finish	Description
8300	8300-N	No. 3 Chain 3/32 in. dia. balls
8302-1	8302-1N	No. 6 Chain, 1/8 in. dia. balls
8373	8373-N	Connector for No. 3 Chain
9820	9820-N	Connector for No. 6 Chain
8376	8376-N	End Bell for Nos. 3 and 6 Chain
8375	8375-N	Oversized Ball (5/16 in. O.D. closed) for hanging fixture bowls with No. 6 Chain
9839	9839-N	Insulating Link, phenol fiber, 1¾ in. total length
6	6-N	Chain, 3 ft. long, end bell and connector; use as extension or replacement
	8010	Cord (white braided), 3 ft. long, end bell and connector; use as extension or replacement

15 AMP

LEVITON J3

INCANDESCENT LAMPHOLDERS

INCANDESCENT LAMPHOLDERS

The following lampholders are designed to meet a wide variety of applications. Leviton's extensive engineering and manufacturing facilities can provide custom designed devices for special applications. For complete information, consult your local Leviton Sales Representative. ALL DIMENSIONS IN THIS SECTION ARE IN INCHES

METAL SHELL LAMPHOLDERS

Refer to Technical Section for typical dimensions of Electrolier, Short Electrolier, and Standard Metal Shell Lampholders.

MEDIUM BASE FINISHES

A suffix must be added to each basic catalog number to indicate the finish desired. For Polished Gilt/Brass finish, add -PG; for Unfinished Aluminum, add -AL; for Polished Aluminum, add -NI (finishes on aluminum). For Bright dip/Brass, add -BR (Finish on brass). CONSULT PRICE LIST FOR AVAILABILITY OF FINISHES.

8002

9346

10083-16

6098

8004

501

STANDARD GRADE
These general duty devices are backed by a Limited Two-Year Warranty.

Cat. No.	Rating	Description
8002	250W 250V	Pull chain—standard. Single circuit. ⅛ IPS tapped bushing w/set screw.
9346	250W 250V	Key—Electrolier. Single Circuit. ⅛ IPS tapped bushing w/set screw.
10083-16	250W 250V	Removable turn knob—electrolier single circuit. ⅛ IPS tapped bushing w/set screw.
7070	250W 250V	Removable turn knob—electrolier 2-circuit, 3 terminal. ⅛ IPS tapped bushing less set screw.
7090	250W 250V	Removable turn knob—electrolier. 2-circuit. ⅛ IPS tapped bushing.
9842	250W 250V	Key-standard. Single circuit. ⅛ IPS tapped bushing w/set screw.
19980	660W 250V	Pull chain—electrolier. Single circuit. ⅛ IPS tapped bushing w/set screw.
6098	660W 250V	Push through—electrolier. Single circuit. ⅛ IPS tapped bushing w/set screw.
8004	660W 250V	Keyless—electrolier. Single circuit. ⅛ IPS tapped bushing w/set screw.
9347	660W 250V	Keyless—short electrolier. Single circuit. ⅛ IPS tapped bushing w/set screw.
9843	660W 250V	Keyless—standard. Single circuit. ⅛ IPS tapped bushing, less set screw.
INTERMEDIATE BASE (same finishes as above)		
512-3	75W 125V	Non-removable turn knob. Single circuit. ⅛ IPS tapped bushing w/set screw.
501	75W 250V	Keyless—single circuit. ⅛ tapped bushing w/set screw.

All items above are UL Listed, except where noted.

J4

15 AMP INCANDESCENT LAMPHOLDERS

19980-M

97092

95070

95110

95080

METAL SHELL LAMPHOLDERS—INTERIORS

	Lampholder	Cat. No. Interior	Lampholder	Cat. No. Interior
STANDARD GRADE ▪ Medium Base ▪ Phenolic body ▪ Aluminum screw-shell.	8002	8002-M	9842	9842-M
	19980	19980-M	9346	9346-M
	10083-16	10083-M	6098	7080-M
	7070	7070-M	8004	8004-M
	7090	7090-M	9843	9843-M

SPECIAL EXTENSION ASSEMBLY

Special Extension Assembly

1⁹⁄₁₆"

No. X-22994 Extension:

Adapter   See list at right

6-32

For applications where standard mandrel is too short

NOTE: Knob and extension shaft must be crimped

Cat. No.	Description
X-22994	Extension Adapter—⁹⁄₁₆ in.
B-1074	Extension Shaft—¾ in.
B-9800	Extension Shaft—1 in.
B-9801	Extension Shaft—1¾ in.
B-9802	Extension Shaft—⁷⁄₁₆ in.
B-9803	Extension Shaft—1¼ in.
BA-5276A	Phenolic Knob with 6-32 Tapped Metal Insert—11/16 in.

BROWN PHENOLIC CAP AND SHELL LAMPHOLDERS

	Cat. No.	Rating	Description	Interiors
STANDARD GRADE *These general duty devices are backed by a Limited Two-Year Warranty.* ▪ Two-piece, medium base electrolier interiors.	97092	250W 250V	Pull cord (nylon), 2-circuit. ⅛ IPS threaded cap w/set screw.	7092-2M 2-circuit
	95070*	250W 250V	Key-type, ⅛ IPS threaded cap w/set screw.	9346-M
	95073	250W 250V	Key-type. Pendant cap w/⁵⁄₁₆ smooth hole.	9346-M
	95110	660W 250V	Push-through; ⅛ IPS threaded cap w/set screw.	7080-M
	95080	660W 250V	Keyless, ⅛ IPS threaded cap w/set screw.	8004-M
	95083*	660W 250V	Keyless, Pendant cap w/⁵⁄₁₆ smooth hole.	8004-M

All items above are UL Listed, except where noted.   *When existing stock is depleted, this item will be DISCONTINUED

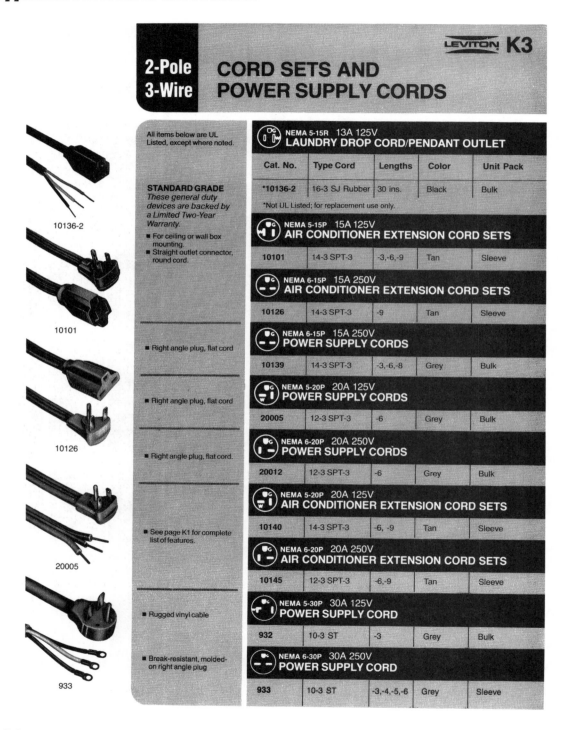

2-Pole 3-Wire CORD SETS AND POWER SUPPLY CORDS

LEVITON K3

Images (left column):
- 10136-2
- 10101
- 10126
- 20005
- 933

All items below are UL Listed, except where noted.

STANDARD GRADE
These general duty devices are backed by a Limited Two-Year Warranty.
- For ceiling or wall box mounting.
- Straight outlet connector, round cord.

- Right angle plug, flat cord

- Right angle plug, flat cord

- Right angle plug, flat cord.

- See page K1 for complete list of features.

- Rugged vinyl cable

- Break-resistant, molded-on right angle plug

NEMA 5-15R 13A 125V — LAUNDRY DROP CORD/PENDANT OUTLET

Cat. No.	Type Cord	Lengths	Color	Unit Pack
*10136-2	16-3 SJ Rubber	30 ins.	Black	Bulk

*Not UL Listed; for replacement use only.

NEMA 5-15P 15A 125V — AIR CONDITIONER EXTENSION CORD SETS

Cat. No.	Type Cord	Lengths	Color	Unit Pack
10101	14-3 SPT-3	-3,-6,-9	Tan	Sleeve

NEMA 6-15P 15A 250V — AIR CONDITIONER EXTENSION CORD SETS

Cat. No.	Type Cord	Lengths	Color	Unit Pack
10126	14-3 SPT-3	-9	Tan	Sleeve

NEMA 6-15P 15A 250V — POWER SUPPLY CORDS

Cat. No.	Type Cord	Lengths	Color	Unit Pack
10139	14-3 SPT-3	-3,-6,-8	Grey	Bulk

NEMA 5-20P 20A 125V — POWER SUPPLY CORDS

Cat. No.	Type Cord	Lengths	Color	Unit Pack
20005	12-3 SPT-3	-6	Grey	Bulk

NEMA 6-20P 20A 250V — POWER SUPPLY CORDS

Cat. No.	Type Cord	Lengths	Color	Unit Pack
20012	12-3 SPT-3	-6	Grey	Bulk

NEMA 5-20P 20A 125V — AIR CONDITIONER EXTENSION CORD SETS

Cat. No.	Type Cord	Lengths	Color	Unit Pack
10140	14-3 SPT-3	-6, -9	Tan	Sleeve

NEMA 6-20P 20A 250V — AIR CONDITIONER EXTENSION CORD SETS

Cat. No.	Type Cord	Lengths	Color	Unit Pack
10145	12-3 SPT-3	-6,-9	Tan	Sleeve

NEMA 5-30P 30A 125V — POWER SUPPLY CORD

Cat. No.	Type Cord	Lengths	Color	Unit Pack
932	10-3 ST	-3	Grey	Bulk

NEMA 6-30P 30A 250V — POWER SUPPLY CORD

Cat. No.	Type Cord	Lengths	Color	Unit Pack
933	10-3 ST	-3,-4,-5,-6	Grey	Sleeve

K4

3-Pole 3 & 4 Wire

CORD SETS AND POWER SUPPLY CORDS

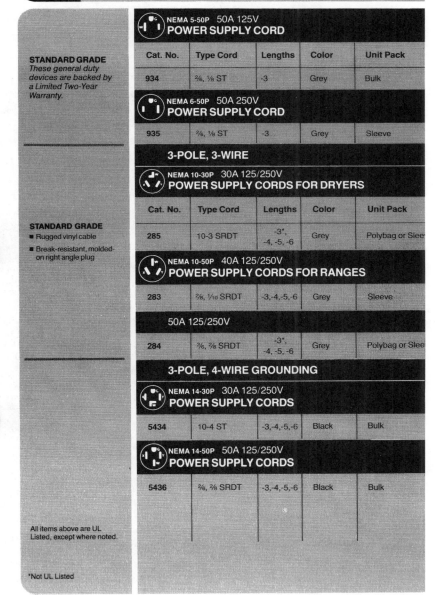

STANDARD GRADE
These general duty devices are backed by a Limited Two-Year Warranty.

NEMA 5-50P 50A 125V
POWER SUPPLY CORD

Cat. No.	Type Cord	Lengths	Color	Unit Pack
934	⅔, ⅛ ST	-3	Grey	Bulk

NEMA 6-50P 50A 250V
POWER SUPPLY CORD

| 935 | ⅔, ⅛ ST | -3 | Grey | Sleeve |

3-POLE, 3-WIRE

NEMA 10-30P 30A 125/250V
POWER SUPPLY CORDS FOR DRYERS

STANDARD GRADE
■ Rugged vinyl cable
■ Break-resistant, molded-on right angle plug

Cat. No.	Type Cord	Lengths	Color	Unit Pack
285	10-3 SRDT	-3*, -4, -5, -6	Grey	Polybag or Slee

285

NEMA 10-50P 40A 125/250V
POWER SUPPLY CORDS FOR RANGES

| 283 | ⅔, ⅒ SRDT | -3,-4,-5,-6 | Grey | Sleeve |

50A 125/250V

| 284 | ⅔, ⅔ SRDT | -3*, -4, -5, -6 | Grey | Polybag or Slee |

3-POLE, 4-WIRE GROUNDING

NEMA 14-30P 30A 125/250V
POWER SUPPLY CORDS

| 5434 | 10-4 ST | -3,-4,-5,-6 | Black | Bulk |

5434

NEMA 14-50P 50A 125/250V
POWER SUPPLY CORDS

| 5436 | ⅔, ⅔ SRDT | -3,-4,-5,-6 | Black | Bulk |

All items above are UL Listed, except where noted.

*Not UL Listed

6700-15

6700-20

6710-15

6710-20

58015-SA

All items below are UL Listed, except where noted.

STANDARD GRADE
Knife switches are backed by a Limited Two-Year Warranty. Fuses are not warrantied.

STANDARD GRADE
Knife switches are backed by a Limited Two-Year Warranty. Fuses are not warrantied.

 LEVITON O1

FUSES AND KNIFE SWITCHES

GLASS PLUG FUSES

■ Escape vent for hot gases caused by short circuit prevents dangerous arcing.
■ Amp rating is visible through window on color coded insert.
■ Meets Fed Specs: W-F831, Type 1; W-F7918, Type 11, Style A, Class 1; MIL-F-15160, Style F14, Characteristic A.

125V AC Only

Cat. No.	AMPS	Insert Color	Window Shape
6700-10	10	Yellow	Hexagonal
6700-15	15	Blue	Hexagonal
6700-20	20	Orange	Round
6700-25	25	Red	Round
6700-30	30	Green	Round

TAMP-PRUF™ TIME DELAY TYPE "S" FUSES

■ Special thermal cut-out provides time-delay, permitting normal current surges (like motor and appliance start-ups) without needless blowing—in accordance with limits set by UL Standards.
■ TAMP-PRUF™ Type "S" design prevents by-pass tampering, for example, insertion of a coin-shaped slug.
■ Improperly rated fuses cannot be substituted, as each fuse will fit only its proper fuseholder or adapter.
■ Adapter screws into standard Edison-base fuseholder or socket; locks in place so it cannot be moved, and only the proper type "S" fuse will fit.
■ Cap is color coded for amp rating, which is printed next to window.

TYPE "S" FUSES
125V AC Only

Cat. No.	AMPS	Cap Color	Window Shape
6710-10	10	Brown	Hexagonal
6710-15	15	Blue	Hexagonal
6710-20	20	Orange	Round
6710-25	25	Green	Round
6710-30	30	Green	Round

TYPE "S" ADAPTERS

Cat. No.	For Fuse Size
58015-SA	10A, 15A
58020-SA	20A
58030-SA	25A, 30A

WALL JACKS   **P1**

Housing Material: High impact, self-extinguishing ABS plastic. Meets UL standard 94-VO rating
Spring Wire Contacts: .000050 in. hard gold plated on phosphor bronze base
Wire Leads: Polyester insulated 26 AWG stranded copper
Terminals: Brass
Screws: Nickel plated steel, shoulder headed
Washers: Nickel plated steel

CAT. NO./COLOR	DESCRIPTION
40249 — Brown **40249-I** — Ivory **40249-W** — White	**TYPE 625 MODULAR JACK FLUSH MOUNT** 4-conductor—One piece construction, installs easily in standard electrical box. Includes jack, plate, mounting screws. For use with 4-conductor line cords. Plate dimensions—2¾ in. W x 4½ in. H x ³/₁₆ in. D overall. Complies with USOC Codes RJ16X and RJ1DC.
40249*-CI	Same as above, carded.
40238 -I — Ivory	**TYPE 625 MODULAR JACK FLUSH MOUNT** 6-conductor—One piece construction, installs easily in standard electrical box. Includes jack, plate, mounting screws. For use with 6-conductor or 4-conductor modular line cords. Plate dimensions—2¾ in. W x 4½ in. H x 3¹/₁₆ in. D overall. Complies with USOC Codes RJ25C.
40549 — Brown **40549-I** — Ivory **40549-W** — White *NEW*	**MID-WAY TYPE 625 MODULAR JACK FLUSH MOUNT** 4-conductor—One piece construction, installs easily in standard electrical box. Includes jack, mounting screws. For use with 4-conductor line cords. Complies with USOC Codes RJ11C, RJ12C, RJ13C, RJ14C, RJ16X, RJ1DC. Mid-Way Plates are ³/₈ in. higher and ³/₈ in. wider than standard plates.
40538 — Brown **40538-I** — Ivory *NEW*	**MID-WAY TYPE 625 MODULAR JACK FLUSH MOUNT** 6-conductor—One piece construction, installs easily in standard electrical box. Includes jack, mounting screws. For use with 6-conductor or 4-conductor modular line cords. Complies with USOC Codes RJ25C. Mid-Way Plates are ³/₈ in. higher and ³/₈ in. wider than standard plates.
40254 — Brown **40254-I** — Ivory **40254-W** — White	**TYPE 625B3 DUPLEX MODULAR JACK** 4-conductor—One piece construction, installs easily in standard electrical box. Includes 2 jacks, plate, mounting screws. Use for dual line wiring phone/answering machine combination. Can be wired to provide same or individual dial tone service. For use with 4-conductor modular line cords. Complies with USOC Codes RJ11C, RJ12C, RJ13C, RJ14C, RJ16X, RJ1DC.
40254-CI*	Same as above, carded.

*Specification compliance with FCC Part 68 requirements is assured only when this item is packaged in standard Leviton boxes.

Appendix: Procedures and Products

P2 LEVITON

CAT. NO./COLOR	DESCRIPTION
40649 — Brown **40649-I** — Ivory **40649-W** — White *NEW*	**DECORA STYLE TYPE 625 MODULAR JACK FLUSH MOUNT** 4-conductor—Can be mounted in any standard electrical box when used in conjunction with Decora wallplate (not included). For use with 4-conductor modular line cords. Several units may be ganged together in a multiple gang box. Complies with USOC Codes RJ11C, RJ12C, RJ13C, RJ14C, RJ16X, RJ1DC.
40638 — Brown **40638-I** — Ivory *NEW*	**DECORA STYLE TYPE 625 MODULAR JACK FLUSH MOUNT** 6-conductor—Can be mounted in any standard electrical box when used in conjunction with Decora wallplate (not included). For use with 6-conductor and 4-conductor modular line cords. Several units may be ganged together in a multiple gang box. Complies with USOC Codes RJ25C.
80401 — Brown **80401-I** — Ivory **80401-W** — White	**DECORA WALLPLATES FOR CAT. NOS. 40649 AND 40638—** Single device. UL Listed. Plastic. Plate dimensions—2¾ in. W x 3⅞ in. H x ¼ in. D overall.
80409 — Brown **80409-I** — Ivory **80409-W** — White	**DECORA WALLPLATES FOR CAT. NOS. 40649 AND 40638—** Two devices. UL Listed. Plastic.
40229-I — Ivory	**TYPE 625B2 MODULAR JACK FLUSH MOUNT ROUND** 4-conductor—One piece construction, installs easily in TYPE 63A or 63C junction box. Includes jack plate, ring and mounting screws. For use with 4-conductor modular line cords. Complies with USOC Codes RJ11C, RJ12C, RJ13C, RJ14C, RJ16X, RJ1DC.

LEVITON® **P3**

CAT. NO./COLOR	DESCRIPTION
40214-I — Ivory NEW	**TYPE 630 MODULAR WALL PHONE JACK WITH PLASTIC FACEPLATE** 4-conductor (screw terminals)—Can be back wired and mounted in any standard electrical box or surface wired and mounted onto plasterboard. Includes snap-on plastic faceplate and factory installed lugs. For use with 4-contact modular wall mount phones. Complies with USOC Codes RJ11C, RJ11W, RJ12C, RJ12W, RJ13C, RJ13W, RJ14C, RJ14W, RJ16X, RJ1DC.
40216-I — Ivory NEW	**TYPE 630 MODULAR WALL PHONE JACK WITH PLASTIC FACEPLATE** 6-conductor (screw terminals)—Can be back wired and mounted in any standard electrical box or surface wired and mounted onto plasterboard. Includes snap-on plastic faceplate and factory installed lugs. For use with 6-contact modular wall mount phones. Complies with USOC Codes RJ25C.
40223-S	**TYPE 630 MODULAR WALL PHONE JACK WITH STAINLESS STEEL FACEPLATE** 4-conductor (quick connect type terminals)—Can be back wired and mounted in standard electrical box or surface wired and mounted on plasterboard. Includes jack, plate, stainless steel faceplate, mounting screws and installation tool. For use with 4-contact modular wall mount phones. Complies with USOC Codes RJ11C, RJ11W, RJ12C, RJ12W, RJ13C, RJ13W, RJ14C, RJ14W, RJ16X, RJ1DC.
40226-S NEW	**TYPE 630 MODULAR WALL PHONE JACK WITH STAINLESS STEEL FACEPLATE** 6-conductor (screw terminals)—Can be back wired and mounted in any standard electrical box or surface wired and mounted on plaster board. Includes jack, plate, stainless steel faceplate, mounting screws. For use with 6-contact modular phones. Complies with USOC Codes RJ25C.
40253-I — Ivory 40253*-CI	**TYPE 630 MODULAR WALL PHONE JACK WITH PLASTIC FACEPLATE** 4-conductor (quick connect terminals)—Can be back wired and mounted in any standard electrical box or surface wired and mounted on plasterboard. Includes plastic faceplate, mounting lugs, plastic color keyed insert and installation tool. For use with 4-contact modular phones. Complies with USOC Codes RJ11C, RJ11W, RJ12C, RJ12W, RJ13C, RJ13W, RJ14C, RJ14W, RJ16X, RJ1DC. Same as above, carded.
40263-I — Ivory	**TYPE 630 MODULAR WALL PHONE JACK WITH PLASTIC FACEPLATE** 6-conductor (quick connect terminals)—Can be back wired mounted in any standard electrical box or surface wired and mounted on plasterboard. Includes plastic faceplate, mounting lugs, plastic color keyed inserts, and installation tool. For use with 4-contact modular wall mount phones. Complies with USOC Codes RJ25C.

*Specification compliance with FCC Part 68 requirements is assured only when this item is packaged in standard Leviton boxes.

P4 LEVITON

CAT. NO./COLOR	DESCRIPTION
**40221-I — Ivory	**TYPE 625 MODULAR JACK FLUSH MOUNT ROUND** 4-conductor— Can be mounted in any electrical box when used in conjunction with TYPE 43A mounting bracket and TYPE 19A faceplate. For use with 4-conductor modular line cords. Complies with USOC Codes RJ11C, RJ12C, RJ13C, RJ14C, RJ16X, RJ1DC.
40201-I — Ivory **NEW**	**TYPE 625 MODULAR JACK FLUSH MOUNT ROUND** 4-conductor— Same as Cat. No. 40221-I above, with TYPE 43A mounting bracket attached. Mounts in any standard electrical box and can be used with an outdoor cover such as Leviton Cat. No. 4925 or equivalent or a TYPE 19A faceplate for indoor use. For use with 4-conductor modular cord. Complies with USOC Codes RJ11C, RJ12C, RJ13C, RJ14C, RJ16X, RJ1DC.
4925 — Grey	**FLUSH MOUNT OUTDOOR WALLPLATE**—Corrosion resistant baked-on grey enamel finish. For use with surface mounted outdoor-type outlet boxes or flush mounted wall boxes. Mounting screws and rubber gasket included.
**40230-5	**MODULAR JACK OUTDOOR FLUSH MOUNT** 4-conductor—Consists of TYPE 625 jack, ring, TYPE 43A mounting bracket, enameled brass faceplate, weather resistant neoprene gaskets and spring loaded cover. For use with 4-conductor modular line cords. Complies with USOC Codes RJ11C, RJ12C, RJ13C, RJ14C, RJ16X, RJ1DC.
40256-CI* — Ivory	**TYPE 496A CONVENTIONAL JACK FLUSH MOUNT** 4-prong— Mounts in any standard electrical box. Consists of TYPE 44 bracket and mounting screws. For use with standard 4-prong plug ended line cords. Carded only.

*Specification compliance with FCC Part 68 requirements is assured only when this item is packaged in standard Leviton boxes.

** When existing stock is depleted, this item will be DISCONTINUED.

SURFACE MOUNTED JACKS

LEVITON **P5**

Housing Material: High impact, self-extinguishing ABS plastic, meets UL standard 94 V-O rating
Spring Wire Contacts: .000050 in. hard gold plated on phosphor bronze base
Wire Leads: Polyester insulated 26 AWG stranded copper
Terminal Lugs: Brass
Screws: Nickel plated steel, shoulder headed
Washers: Nickel plated steel

	CAT. NO./COLOR	DESCRIPTION
	40274 — Brown 40274-I — Ivory **NEW**	**TYPE 625 MODULAR JACK SURFACE MOUNT** 4-conductor—Consists of snap-on cover and base with pre-wired modular jack. For use with 4-conductor modular line cords. Complies with USOC Codes RJ11C, RJ12C, RJ13C, RJ14C, RJ16X, RJ1DC.
	40276-I — Ivory **NEW**	**TYPE 625 MODULAR JACK SURFACE MOUNT** 6-conductor—Consists of snap-on cover and base with pre-wired modular jack. For use with 6-conductor or 4-conductor modular line cords. Complies with USOC Codes RJ25C.
	40278-I — Ivory **NEW**	**TYPE 625 MODULAR JACK SURFACE MOUNT** 8-conductor—Consists of a snap-on cover and base with pre-wired modular jack. For use with 8-conductor non-keyed modular line cords.
	40278-SBI — Ivory **NEW**	**TYPE 625 MODULAR JACK SURFACE MOUNT** 8-conductor—Consists of snap-on cover and base with pre-wired modular jack incorporating shorting bars. For use with 8-conductor non-keyed modular line cords. Commonly used with burglar and fire alarm equipment that sends emergency electronic signals over the tip and ring dial tone to a central monitoring point. Complies with USOC Codes RJ31X through RJ38X.
	40237-I — Ivory	**TYPE 625 MODULAR JACK SURFACE MOUNT** 4-conductor—Consists of TYPE 625C jack and cover and TYPE 42A terminal block. For use with 4-conductor modular line cords. Special colors available. Consult factory for availability. Complies with USOC Codes RJ11C, RJ12C, RJ13C, RJ14C, RJ16X, RJ25C, RJ1DC.

Appendix: Procedures and Products

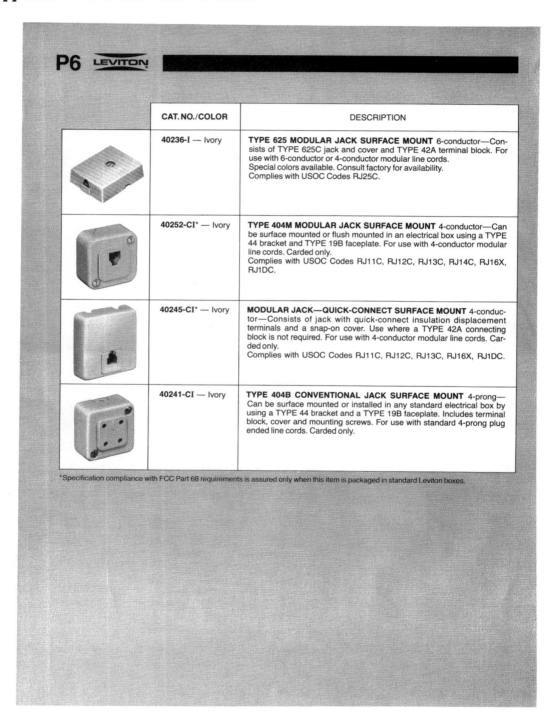

P6 LEVITON

CAT. NO./COLOR	DESCRIPTION
40236-I — Ivory	**TYPE 625 MODULAR JACK SURFACE MOUNT** 6-conductor—Consists of TYPE 625C jack and cover and TYPE 42A terminal block. For use with 6-conductor or 4-conductor modular line cords. Special colors available. Consult factory for availability. Complies with USOC Codes RJ25C.
40252-CI* — Ivory	**TYPE 404M MODULAR JACK SURFACE MOUNT** 4-conductor—Can be surface mounted or flush mounted in an electrical box using a TYPE 44 bracket and TYPE 19B faceplate. For use with 4-conductor modular line cords. Carded only. Complies with USOC Codes RJ11C, RJ12C, RJ13C, RJ14C, RJ16X, RJ1DC.
40245-CI* — Ivory	**MODULAR JACK—QUICK-CONNECT SURFACE MOUNT** 4-conductor—Consists of jack with quick-connect insulation displacement terminals and a snap-on cover. Use where a TYPE 42A connecting block is not required. For use with 4-conductor modular line cords. Carded only. Complies with USOC Codes RJ11C, RJ12C, RJ13C, RJ16X, RJ1DC.
40241-CI — Ivory	**TYPE 404B CONVENTIONAL JACK SURFACE MOUNT** 4-prong—Can be surface mounted or installed in any standard electrical box by using a TYPE 44 bracket and a TYPE 19B faceplate. Includes terminal block, cover and mounting screws. For use with standard 4-prong plug ended line cords. Carded only.

*Specification compliance with FCC Part 68 requirements is assured only when this item is packaged in standard Leviton boxes.

272

JUNCTION BLOCKS

LEVITON **P7**

Housing Material: High impact, self-extinguishing ABS plastic, meets UL standard 94-VO rating
Modular Plug: High impact polycarbonate
Plug Contacts: .000050 in. hard gold plated on phosphor bronze base (40218 only)
Wire Leads: Polyester insulated 26 AWG stranded copper
Insulation Displacement Terminals: Tin plated over brass
Screws: Nickel plated steel, oversize head
Washers: Nickel plated steel

CAT. NO./COLOR	DESCRIPTION
40218-I — Ivory	**TYPE 742A WIRE JUNCTION BLOCK WITH MODULAR PLUG** 4-conductor—Plug connects to telephone utility demarcation jack. Junction block acts as a starting point for inside phone system wiring. Includes cover, lid, base, 4-conductor cord with modular plug end and mounting screws.
40219-I — Ivory	**TYPE 742B WIRE JUNCTION BLOCK** 4-conductor—Junction point for inside phone system wiring. Includes cover, lid, base and mounting screws.

MODULAR LINE CORD PLUG AND CRIMPING TOOL

Modular Plug: High impact polycarbonate
Contacts: .000050 in. hard gold plated on phosphor bronze base

CAT. NO./COLOR	DESCRIPTION
49404	**MODULAR LINE CORD PLUG**—6-position, 4-conductor for flat or oval cable. For conductors with maximum outer diameter of 0.038 in. Can be attached to flat modular line cordage (Cat. No. 42404-5CS) with Crimping Tool (Cat. No. 49400) to make custom length modular line cords.
49400	**CRIMPING TOOL**—Heavy duty, for installing 6-position, 4-conductor line cord plugs (Cat. No. 49404 above) on 4-conductor line cordage (Cat. No. 42404-5CS). Wire cutter, crimper and stripper built in.

NEW

P8 PLUGS—ADAPTERS—COUPLERS

Housing Material: High impact ABS plastic
Contacts: .000020 in. hard gold plated on phosphor bronze base
Wire Leads: Polyester insulated 26 AWG stranded copper
Terminals: Brass

CAT. NO./COLOR	DESCRIPTION
40246-CI* — Ivory	**TYPE 225A CONVERSION ADAPTER** 4-prong to 4-contact modular. Plugs into a standard 4-prong jack converting it for use with 4-conductor modular line cords. Carded only. Complies with USOC Code RJA1X.
40247-CI* — Ivory	**TYPE 267A MODULAR DUPLEX "Y" ADAPTER**—Converts single 4-contact modular outlet to two 4-contact modular outlets. For use with 4-conductor modular line cords. Carded only. Complies with USOC Code RJA2X.
40250-CI* — Ivory	**MODULAR CORD COUPLER** 4-contact—For use with 4-conductor modular line cords. Allows connection to two modular cords. Carded only.
40251-CI* — Ivory	**RETROFIT CONVERSION PLUG** 4-conductor modular—Converts 4-conductor ring- or spade-tipped line cords for use with modular jacks. Includes plug and cover. Carded only.
40240-CI* — Ivory	**TYPE 283B CONVENTIONAL PLUG** 4-prong—Can be mounted on end of 4-conductor line cord for compatibility with standard 4-prong jacks TYPE 498A, 550A, 404B and 548A. Includes plug and cover. Carded only.

*Specification compliance with FCC Part 68 requirements is assured only when this item is packaged in standard Leviton boxes.

T3 LEVITON® ANSI Architectural Symbols

1. Lighting Outlets

	Ceiling	*Wall*

1.1 Surface or Pendant Incandescent, Mercury-Vapor, or Similar Lamp Fixture

1.2 Recessed Incandescent, Mercury-Vapor, or Similar Lamp Fixture

1.3 Surface or Pendant Individual Fluorescent Fixture

1.4 Recessed Individual Fluorescent Fixture

1.5 Surface or Pendant Continuous-Row Fluorescent Fixture

1.6 Recessed Continuous-Row Fluorescent Fixture

1.7 Bare-Lamp Fluorescent Strip

1.8 Surface or Pendant Exit Light

1.9 Recessed Exit Light

1.10 Blanked Outlet

1.11 Junction Box

1.12 Outlet Controlled by Low-Voltage Switching When Relay Is Installed in Outlet Box

2. Receptacle Outlets

	Grounded	*Ungrounded*

2.1 Single Receptacle Outlet

2.2 Duplex Receptacle Outlet

2.3 Triplex Receptacle Outlet

2.4 Quadruplex Receptacle Outlet

2.5 Duplex Receptacle Outlet—Split Wired

2.6 Triplex Receptacle Outlet—Split Wired

2.7 Single Special-Purpose Receptacle Outlet

2.8 Duplex Special-Purpose Receptacle Outlet

2.9 Range Outlet (typical)

2.10 Special-Purpose Connection or Provision for Connection

	Grounded	*Ungrounded*

2.11 Multioutlet Assembly

2.12 Clock Hanger Receptacle

2.13 Fan Hanger Receptacle

2.14 Floor Single Receptacle Outlet

2.15 Floor Duplex Receptacle Outlet

2.16 Floor Special-Purpose Outlet

3. Switch Outlets

3.1 Single-Pole Switch **S**

3.2 Double-Pole Switch **S2**

3.3 Three-Way Switch **S3**

3.4 Four-Way Switch **S4**

3.5 Key-Operated Switch **SK**

3.6 Switch and Pilot Lamp **SP**

3.7 Switch for Low-Voltage Switching System **SL**

3.8 Master Switch for Low-Voltage Switching System **SLM**

3.9 Switch and Single Receptacle **S**

3.10 Switch and Double Receptacle **S**

3.11 Door Switch **SD**

3.12 Time Switch **ST**

3.13 Circuit Breaker Switch **SCB**

3.14 Momentary Contact Switch or Pushbutton for Other Than Signaling System **SMC**

3.15 Ceiling Pull Switch (S)

5. Residential Occupancies

5.1 Pushbutton

5.2 Buzzer

5.3 Bell

5.4 Combination Bell–Buzzer

5.5 Chime

5.6 Annunciator

NEMA
Straight Blade Configurations — LEVITON T4

NEMA CONFIGURATIONS FOR GENERAL-PURPOSE NONLOCKING PLUGS AND RECEPTACLES

		#	15 AMPERE RECEPTACLE	15 AMPERE PLUG	20 AMPERE RECEPTACLE	20 AMPERE PLUG	30 AMPERE RECEPTACLE	30 AMPERE PLUG	50 AMPERE RECEPTACLE	50 AMPERE PLUG	60 AMPERE RECEPTACLE	60 AMPERE PLUG
2-POLE 2-WIRE	125V	1	1-15R	1-15P								
	250V	2		2-15P	2-20R	2-20P	2-30R	2-30P				
	277V	3	(RESERVED FOR FUTURE CONFIGURATIONS)									
	600V	4	(RESERVED FOR FUTURE CONFIGURATIONS)									
2-POLE 3-WIRE GROUNDING	125V	5	5-15R	5-15P	5-20R	5-20P	5-30R	5-30P	5-50R	5-50P		
	250 V	6	6-15R	6-15P	6-20R	6-20P	6-30R	6-30P	6-50R	6-50P		
	277V AC	7	7-15R	7-15P	7-20R	7-20P	7-30R	7-30P	7-50R	7-50P		
	347V AC	24	24-15R	24-15P	24-20R	24-20P	24-30R	24-30P	24-50R	24-50P		
	480V AC	8	(RESERVED FOR FUTURE CONFIGURATIONS)									
	600V AC	9	(RESERVED FOR FUTURE CONFIGURATIONS)									
3-POLE 3-WIRE	125/250V AC	10			10-20R	10-20P	10-30R	10-30P	10-50R	10-50P		
	3∅ 250 V	11	11-15R	11-15P	11-20R	11-20P	11-30R	11-30P	11-50R	11-50P		
	3∅ 480V	12	(RESERVED FOR FUTURE CONFIGURATIONS)									
	3∅ 600V	13	(RESERVED FOR FUTURE CONFIGURATIONS)									
3-POLE 4-WIRE GROUNDING	125/250V	14	14-15R	14-15P	14-20R	14-20P	14-30R	14-30P	14-50R	14-50P	14-60R	14-60P
	3∅ 250V	15	15-15R	15-15P	15-20R	15-20P	15-30R	15-30P	15-50R	15-50P	15-60R	15-60P
	3∅ 480V	16	(RESERVED FOR FUTURE CONFIGURATIONS)									
	3∅ 600V	17	(RESERVED FOR FUTURE CONFIGURATIONS)									
4-POLE 4-WIRE	3∅ 208Y/120V	18	18-15R	18-15P	18-20R	18-20P	18-30R	18-30P	18-50R	18-50P	18-60R	18-60P
	3∅ 480Y/277V	19	(RESERVED FOR FUTURE CONFIGURATIONS)									
	3∅ 600Y/347V	20	(RESERVED FOR FUTURE CONFIGURATIONS)									
4-POLE 5-WIRE GROUNDING	3∅ 208Y/120V	21	(RESERVED FOR FUTURE CONFIGURATIONS)									
	3∅ 480Y/277V	22	(RESERVED FOR FUTURE CONFIGURATIONS)									
	3∅ 600Y/347V	23	(RESERVED FOR FUTURE CONFIGURATIONS)									

Receptacles

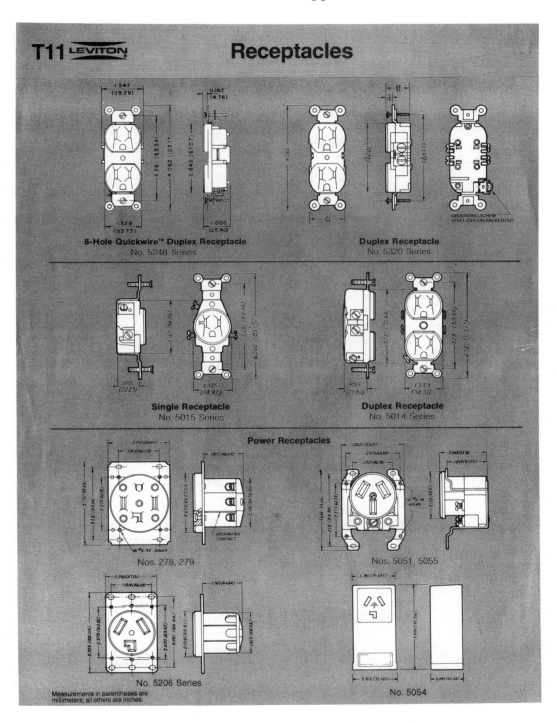

8-Hole Quickwire™ Duplex Receptacle
No. 5248 Series

Duplex Receptacle
No. 5320 Series

GROUNDING SCREW
STEEL (GREEN) BACKED OUT

Single Receptacle
No. 5015 Series

Duplex Receptacle
No. 5014 Series

Power Receptacles

Nos. 278, 279

Nos. 5051, 5055

No. 5206 Series

No. 5054

Measurements in parentheses are
millimeters; all others are inches.

Telephone Wiring Devices
Codes and Standards

Local Codes and Requirements

LOCAL rules and regulations regarding telephone wiring systems must be investigated and followed by the contractor. Even though the Federal Communications Commission (FCC) acted in April, 1984, to replace local control on residential telephone wiring systems with a federal standard to be applied under FCC Part 68, you *must* still investigate and abide by local rules and practices until national standards are fully implemented.

There is, at this writing, no comprehensive national code for telephone wiring systems like the National Electric Code. Some NEC requirements do apply to telephone wiring systems, and are enforced in some jurisdictions, but comprehensive national standards governing all aspects are developing rapidly following general deregulation and growth of competition. UL Listing standards are anticipated by many industry analysts as new regulations take effect.

Until these national standards take effect, each local telephone operating company (telco) will have their own standards and requirements. These are usually developed under tariffs approved by a state utility regulatory commission. The contractor must stay informed on these requirements because they will continue to change as telecommunications deregulation continues. Do not assume that rules in one area will be the same as the rules of a nearby telco in another area. Often, this is not the case. Check with EACH utility company that supplies the dial tone in ALL of the areas where you will be working.

Commercial and Residential Systems

Technological advances in the design of commercial telephone systems may cause them to differ substantially from residential telephone systems. However, many modern electronic key telephone systems (KTS) and branch and private branch exchanges (PBX) are installed in commercial property using many of the same, or similar, materials and methods for the installation of residential systems.

SOME OF THE CONCEPTS AND REGULATIONS DISCUSSED HERE MAY APPLY TO TELEPHONE WIRING SYSTEMS AS WELL AS RESIDENTIAL LOCATIONS. THERE ARE, HOWEVER, OTHER COMPLEX ISSUES AND IMPORTANT RULES TO BE CONSIDERED IN COMMERCIAL TELEPHONE WIRING SYSTEM INSTALLATIONS.

Assistance from the Telephone Utility

The local telephone company will generally be very helpful in assisting the contractor to meet local tariff rules and requirements (at least until adoption of national standards). Most telcos publish a written guide to compliance that is available, free of charge, through the local operating company foreman or Customer Provided Inside Wire (CPIW) coordinator. In most areas, you may reach the foreman or coordinator at the local telco business office to make an appointment to meet him at a particular jobsite. Be sure to ask if the telco will charge for the CPIW person's services. Your distributor or Leviton representative can assist if you are unable to reach someone through the telco.

THERE IS NO AUTOMATIC INSPECTION TO ASSURE COMPLIANCE OF ROUGH-IN TELEPHONE WIRING BEFORE COVER-UP. If you fail to meet the specific requirements of the local telco, they can refuse to connect dial tone service to your wiring system. Many of the problems caused by improper installation are not evident until the dial tone is connected and working. Unless you specifically request assistance from the telco, there will probably not be an inspection of your rough-in work. In most areas, there is no telephone wiring system inspection at all.

Only later, when it is too late to make economical changes, do errors show up. These can be avoided by close cooperation with the telco. Customers will call the telco and complain about interference and "crosstalk" (a second conversation on the line) if improper methods or materials were used by the contractor during wiring. If customer trouble reports are isolated and found to be related to the inside wiring system, the telco will disconnect service until the faulty wiring is repaired or replaced. These repairs can be extremely costly and difficult.

Telecommunications Industry Basics

"Unbundling" is one of many new "telspeak" terms in the emerging deregulated telecommunications industry business language. To "unbundle" telecommunications means to separate, both physically and economically, the various equipment systems and services that work together to provide dial tone and worldwide access from one station to another.

Virtually all telecommunications products and services, except for local exchange NETWORK DIAL TONE, which is still a regulated monopoly, have been unbundled. This process has removed the traditional cross-subsidy situation that served to hide the real cost of providing the CUSTOMER PREMISE EQUIPMENT (CPE), repair, installation, relocation and telephone wiring system work. Most of the cost to provide these services was paid by long-distance callers and monthly recurring "line charges" for local dial tone. No accurate itemized billing was ever provided.

Planning and design services that were once "free" to the architect and engineer are now billed at an hourly rate from the Building Industry Consulting Services (BICS) for the design and specification of telephone wiring systems in commercial buildings.

Hourly and material charges for the installation of these telephone wiring systems are, in most areas, at or above electrical contractor rates for wiring work. Much of the electrical industry, unfortunately, is still unaware of the scope of

Telephone Wiring Devices
Codes and Standards

T23 LEVITON®

telephone wiring work that is now unregulated and open to the traditional wiring industry. When the AT&T divestiture became effective on January 1, 1984, ownership of tens of billions of dollars worth of premise telephone system equipment transferred to the new CPE organization, AT&T INFORMATION SYSTEMS.

Not a penny's worth of the Telephone wiring systems that connect together all the individual pieces of equipment and connect the equipment to the network was transferred to the new AT&T unit. TELEPHONE WIRING SYSTEMS are considered separate from EQUIPMENT. It is the customer's responsibility using the contractor of his choice, to install, maintain, change and connect his own telephone wiring system, whether in a new installation or to accommodate a change in his existing telecommunications requirements.

Separation of Network and Premise

The NETWORK, meaning everything in the telecommunications system that lies outside an individual business or residence, is connected with the PREMISE, the system of all the equipment and material within the building that provides a link between individual telephone sets or other connected equipment, such as computers or answering systems.

There must be a point of separation between the network and premises under new industry regulations. This is called the demarcation point (DEMARC). A subscriber network interface (SNI) is installed at the location where the premises system will connect with the network. This SNI (also called the demarcation jack) is always an easily disconnected plug and jack arrangement that can quickly separate the network and premise systems.

Responsibility for the required new equipment, and system maintenance or repair, is then determined its position in relation to the SNI. If on the network side, it must be provided by the network company. If on the premise side of the SNI, it is the responsibility of the customer.

The Demarcation Point Concept

The telco is only to bring dial tone service (NETWORK) to a certain point (the demarcation jack). At that point, a subscriber network interface (SNI) is installed by the telco. The telephone wiring system and all equipment beyond the SNI is the responsibility of the customer and can be installed by the contractor of choice.

The actual location of the SNI will depend on the local rules. The trend is for the demarcation point to be on the outside of most newer homes, usually within six feet of the electrical meter base. Older homes may have the SNI located at any number of points between the lightning protector (always provided by the telco) and the inside telephone jacks.

Most tariffs will specify different demarcs for different circumstances. Conditions affecting the demarc include number of units, access, type and age of connecting hardware and other considerations. In new residential housing developments, the utility regulatory commission may even permit the SNI to be located at a service point in an easement area some distance from the home. In this case, the homeowner will be responsible for the service drop into the house as well as for the inside wiring.

NETWORK

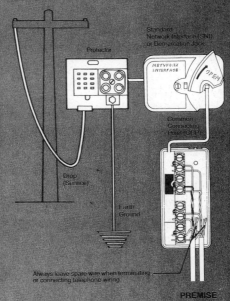

Standard Network Interface (SNI) or Demarcation Jack

Protector

NETWORK INTERFACE   OPEN

Common Connecting Point (CCP)

Drop (Service)

Earth Ground

Always leave spare wire when terminating or connecting telephone wiring.

PREMISE

Minimum Point of Presence (MPOP)

Following deregulation and development of competition for premises wiring, the utility telephone companies are introducing a new doctrine which the regulating utility commissions are rapidly embracing. This doctrine is called Minimum Point of Presence (MPOP). Under MPOP rules, the telephone utility will provide the necessary wiring and materials (under tariff) to bring a dial tone circuit *only* to that point just inside the building where proper terminals and equipment exist. If the customer wants the circuit extended to some other point in the building, the customer must pay time and material charges for that extension of the wiring system and for the installation of the jacks. The customer is NOT to choose the telco for this work. Other contractors may provide wiring beyond the MPOP.

1981 Nonmetallic Box Cubic Capacity Chart

The RACO cubic capacity chart allows electrical specifiers and installers to select nonmetallic outlet and switch boxes with sufficient capacity to meet the requirements of the 1981 National Electrical Code. Section 370-6 of the Code recognizes that conductors, clamps and devices take space in a box and assigns a numerical value to them. Section 370-6 & -7c are reproduced on the back page and should be studied before using this chart. The Code requires the following be counted:

Each current-carrying conductor
Each strap containing one or more devices
Each fixture stud or hickey
One or more ground wires
One or more cable clamps

How to use the chart

Example #1: All conductors the same size.
install two #14-2 nonmetallic sheathed cables with ground plus one duplex receptacle in a box without cable clamps. Think of this as four #14 current carrying conductors and two ground wires. These count as:

4 current-carrying conductors 4
2 ground wires 1
duplex receptacle on a single strap 1
Total number of additions 6

Now read across the "No. of Additions" line in the Box Fill Table to 6 and read down this column to the #14 conductor figure. You need a box without cable clamps with at least 12.00 cubic inch capacity. The chart below lists the RACO boxes that have the capacity.

In this example, if the current carrying conductors were #12, you would read down the 6 column and see that a box with at least 13.50 cubic inches is required. Larger conductors take more space in a box and the Code reflects this fact.

Example #2: Two different size current carrying conductors

When two different size current carrying conductors will be in the same box, the additions for ground wires, clamps and devices must be made for the **larger** conductor, then add the total of the smaller current carrying conductors to it. For example, install two #12-2 and two #14-2 nonmetallic sheathed cables with ground plus two toggle switches (each mounted on its own strap) in a box with cable clamps. These count as:

4 #12 current carrying conductors 4
2 device straps 2
4 ground wires 1
cable clamps 1
Total additions for the larger conductor 8

Read across the "No. of Additions" line in the Box Fill Table to 8 and read down to #12 conductor figure of 18.00. Now add to this the four #14 current carrying conductors of 8.00 for a total box volume of 26.00 cubic-inches.

DESCRIPTION	4" SQUARE 2¼" DEEP	2-GANG SWITCH 3 9/16" x 3⅞" x 2 15/16"	2-GANG SWITCH 3 9/16" x 3⅞" x 2⅝"	4" CEILING 2 9/16" DEEP	4" SQUARE 1⅝" DEEP
CU-IN. CAPACITY	30.0	27.5	22.5	22.5	21.0
typical illustration	 No. 7234	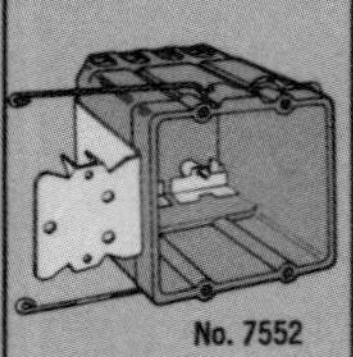 No. 7552	 No. 7491	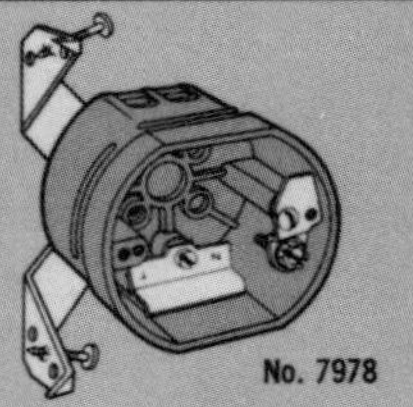 No. 7978	 No. 7221
complete RACO listing	**RACO No. Accessories** with clamps 7234 NP brkt., nails 7230 FP brkt. 7233 BP brkt., set flush without clamps 7244 NP brkt., nails 7240 FP brkt. 7243 BP brkt., set flush	**RACO No. Accessories** with clamps 7546 NP brkt., nails 7545 FP brkt. 7552 BP brkt., ¼" setback without clamps 7544 NP brkt., nails 7543 FP brkt. 7553 BP brkt., ¼" setback	**RACO No. Accessories** with clamps 7490 NP brkt., nails 7491 FP brkt. without clamps 7492 NP brkt., nails 7493 FP brkt.	**RACO No. Accessories** with clamps / without clamps 7178 7177 NP brkt., nail 7978 7977 NP brkt., nails, grd. plate 7175 7176 JP brkt. 7975 7976 JP brkt., grd. plate set-up boxes, adj. bar hanger 7335 7336 11½"-18½" bar 7935 7936 11½"-18½" bar, grd. plate 7338 7337 19½"-26½" bar 7938 7937 19½"-26½" bar, grd. plate	**RACO No. Accessories** with clamps 7218 NP brkt., nails 7221 FP brkt. 7224 BP brkt., set flush without clamps 7219 NP brkt., nails 7220 FP brkt. 7225 BP brkt., set flush

For larger capacities, RACO offers 3-gang switch boxes with 34.0 cubic-inch capacity and 4-gang with 48.0 cubic-inch capacity.

Illustrations on page 280 and 281 are Courtesy Raco Fittings and Boxes.

Raised covers and capacity

RACO offers single and two device covers raised ½" for 4" square nonmetallic boxes. When assembled to the box, they add 4.0 and 6.8 cubic inches respectively allowing additional conductors. In example #2, a box with 26.00 cubic-inches is required. This appears to exceed the 21.0 cubic-inch capacity of the 4" square box 1⅝" deep and require the 2¼" deep box with 30.3 cubic inches. However use of 2-device cover adds 6.8 cubic inches to the box for a total of 27.8, more than enough to permit the use of the less costly 1⅝" deep box.

New thermoplastic "soft" boxes

To offer electricians a choice, RACO is adding 16.0 and 18.0 cubic-inch capacity single gang switch boxes made of "soft" thermoplastic material to our line of thermoset boxes.

Single and two gang thermoset nonmetallic boxes have been accepted by Underwriters' Laboratories, Inc. and the International Conference of Building Officials (ICBO) in 2-hour fire walls. Test report summaries supplied upon request.

BOX FILL TABLE

No. Additions		1	2	3	4	5	6	7	8	9	10	11	12	13	14	15	
CONDUCTOR SIZES	#14	2.00	4.00	6.00	8.00	10.00	12.00	14.00	16.00	18.00	20.00	22.00	24.00	26.00	28.00	30.00	#14
	#12	2.25	4.50	6.75	9.00	11.25	13.50	15.75	18.00	20.25	22.50	24.75	27.00	29.25	31.50	33.75	#12
	#10	2.50	5.00	7.50	10.00	12.50	15.00	17.50	20.00	22.50	25.00	27.50	30.00	32.50	35.00	37.50	#10
	#8	3.00	6.00	9.00	12.00	15.00	18.00	21.00	24.00	27.00	30.00	33.00	36.00	39.00	42.00	45.00	#8
	#6	5.00	10.00	15.00	20.00	25.00	30.00	35.00	40.00	45.00	50.00	55.00	60.00	65.00	70.00	75.00	#6

SWITCH 3½" x 2¼" x 3⅛"

18.0

No. 7301

RACO No.	Accessories
	thermoset
7602	nails
7601	Rammit clips, nails
7605*	FP brkt.
7604*	BP brkt., ¼" setback
7603*	BP brkt., ½" setback
	thermoplastic, 2⅞" deep
7302	nails
7301	Rammit clips, nails

*17.8 cubic inches

SWITCH 3½" x 2¼" x 2⅞"

16.0

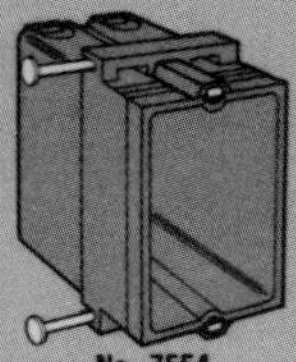

No. 7554

RACO No.	Accessories
	thermoset
7555	nails
7554	Rammit clips, nails
7541	FP brkt.
7540	BP brkt., ¼" setback
7529	BP brkt., ½" setback
	thermoplastic, 2-9/16" deep
7355	nails
7354	Rammit clips, nails

4" CEILING 1⅝" DEEP

14.0

No. 7328

RACO No. with clamps	RACO No. without clamps	Accessories
7163	7164	NP brkt., nails
7963	7964	NP brkt., nails, grd. plate
7161	7160	JP brkt.
7961	7960	JP brkt., grd. plate
set-up boxes — adj. bar hanger		
7328	7329	11½"-18½" bar
7928	7929	11½"-18½" bar, grd. plate
7330	7331	19½"-26½" bar
7930	7931	19½"-26½" bar, grd. plate

3½" ceiling 2⅛" DEEP

14.0

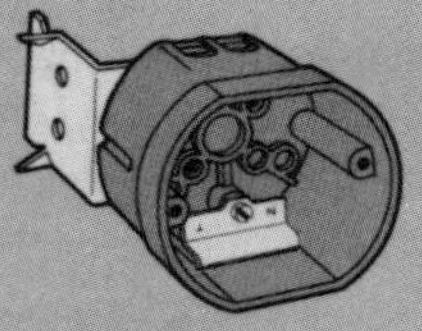

No. 7123

RACO No. with clamps	RACO No. without clamps	Accessories
7118	7119	NP brkt., nails
7918	7919	NP brkt., nails, grd. plate
7123	7120	JP brkt.
7923	7920	JP brkt., grd. plate
7121	——	old work saddle, ears, clamps
7921	——	old work saddle, ears, clamps, grd. plate
set-up boxes, adj. bar hanger		
7305	7304	11½"-18½"bar
7905	7904	11½"-18½" bar grd. plate
7306	7303	19½"-26½" bar
7906	7903	19½"-26½" bar, grd. plate

SWITCH 3½" x 2¼" x 2-3/16"

13.5

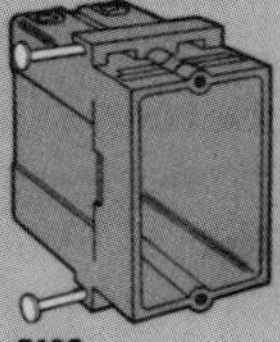

RACO No.	Accessories
7455	nails

SWITCH 3-3/32" x 2-3/16" x 2½"

11.5

RACO No.	Accessories
7487	old work box, saddle, ears, clamps

Glossary

accessible: Can be removed or exposed without damaging any building material, structure, or walls.

ampacity: The current-carrying capacity of electric conductors, expressed in amperes.

appliance: Utilization equipment (current-using) that can either be fixed (permanently wired) or portable (plugged into a receptacle). For example, the electric range, water heater (fixed), clothes dryer, and toaster.

approved: acceptable to the authority having jurisdiction.

attachment plug (plug, cap): Device at the end of an appliance cord to connect an appliance to a receptacle.

automatic: An electric mechanism that starts and stops on its own by means of a controlling device such as a thermostat.

AWG: American Wire Gauge; adopted to standardize wire diameters.

bonding jumper: A wire connection that joins two metal parts that must have electrical contact at all times (removal of a water meter from its place in a water line can disconnect the electrical ground connection from the electrical system).

bonding jumper, main: Jumper that connects the neutral bar in service-entrance panels to the metal of the panel to ground it.

branch circuit: The wires or cable from the fuse or breaker to the point of current usage, as a receptacle or ceiling fixture.

branch circuit, individual: A circuit having only one utilization equipment item connected to it, as an electric range or dryer.

building: A separate structure standing alone or cut off from adjoining structures or by solid walls or fire doors.

bushing, conduit: A screwed fitting put on the end of a conduit has rounded edges to prevent abrasion of wire insulation.

bushing, pipe: A fitting for reducing the size of a pipe thread to accept a smaller pipe or fitting. A fiber bushing inserted in the cut end of ABC (BX) to protect the wire insulation.

BX: Armored Bushed Cable (ABC), similar to Romex in use. Has spirally wrapped steel casing for more protection. Mandatory in some areas.

cabinet: An enclosure that contains wires and devices such as circuit breakers or fuses. Can be surface or flush mounted.

circuit breaker: An overcurrent device that automatically opens the circuit to disconnect it in case of overload or short circuit. Can be switched manually to open the circuit.

close: To operate a switch to complete a circuit. Also called *make*.

concealed: Not accessible without damage to the building or building surface, including conduit and tubing.

conductor: Any wire or busbar that is capable of carrying current; can be bare or insulated.

connector, pressure (solderless): A terminal end that can be attached under a screw or pushed onto or into a mating terminal end. The wire is inserted into the hollow end and is crimped using a special crimping tool to make a secure electrical connection.

continuous load: Any electrical load of over 3 hours duration.

dead: Having no voltage; not attached to a current source.

dead front: A panel or switchgear that has all live parts covered to prevent shock.

device: Any unit carrying current but not utilizing current, as a switch, receptacle, or thermostat.

die: A tool for cutting external threads on a rod or bolt.

dwelling: House, home, living quarters for one or more persons.

Edison-base thread: The thread found on light bulbs and fuses.

electrical bonding: Joining two metal parts to form a solid connection.

elevation: Height above a reference point such as a floor.

feedback (separate feed): A second, separate source of current feeding the same item of equipment. Both sources of current *must* be turned off to safely work on the equipment.

ferrule: A sleeve of brass, used to seal waterproof fittings for flexible conduit.

fill: The percentage of space in an electric box that wires may occupy without crowding. Required by the NEC.

fish: To push or pull cable inside hollow walls or through attics while attempting to snag the end, retrieve it, and pull it through a wall opening. This is classed as *old work*.

fuse: The weak link in a circuit that fails upon overload or short circuit, thereby protecting wires and equipment from damage.

fuse block, pull-out type: Used in service-entrance panels. Consists of a plastic block with fuses, fuse clips, and prongs. The prongs fit into a recessed receptacle that accepts the block, making contact through the fuses to complete the circuit.

fused disconnect: A device with fuses in a cabinet; an externally operated switch (or fuse block) disconnects the power from the equipment.

ground: A connection to earth, whether accidental or intentional; or an object that is connected to earth.

grounded conductor: A conductor that is intentionally grounded.

grounding conductor, equipment: The wire connecting the metal non-current-carrying parts of panels, cabinets, conduit, and motor frames and housings.

hickey: A tool for bending rigid conduit. Similar to a pipe tee with one side cut away.

hot: An energized wire, motor, or other equipment. Also called *live*.

I.D.: Inside diameter, as of a pipe or conduit.

in sight from (within sight from): The ability to see a disconnect that has been turned off from the place where work is being done so the work can be done safely.

knife blade: Refers to a large-amperage fuse having thin blade-type ends instead of round ends. Also refers to disconnects with similar blades.

knockout: Round, partially punched-out openings in cabinets and junction boxes that are knocked or pried out to accept fittings.

lighting outlet: A connection for a lighting fixture, usually in the ceiling, sometimes in a wall.

location, damp: A location sheltered by overhanging eaves or roofs; includes barns and basements.

location, dry: A location not subject to dampness.

location, wet: Underground or in contact with water.

locknut: A flat, thin nut which screws onto a cable or conduit fitting to secure in a knockout.

meter base: A cast or stamped metal box that has connections for the electric meter. The meter plugs into clips inside the box much as an attachment plug is used.

mouse: A handmade tool of coiled wire solder that acts as a weight on the end of a cord. This is lowered inside a hollow wall and fished out through an opening for a receptacle or switch.

nameplate rating: A metal plate stamped with pertinent data. Used on items such as motors, air conditioners, or household appliances. Necessary when ordering replacement parts or calculating current draw.

new work: Electrical work done before the finish wall or ceiling materials are installed. Also called *rough-in work*.

normal: The condition of a switch or other equipment when it has no external force acting on it. It is the state of a device as you hold it in your hand, or as you remove it from its shipping container.

nut driver: A small socket attached to a screw-driver handle for ease of use (rather than a ratchet handle).

O.D.: Outside diameter, as in conduit or tubing.

old work: Installation of electrical wiring in a finished building. Here you must fish the cables and wires inside the walls by cutting and patching. The building might still be new, but the type of work still is called *old*.

open (*noun*): A break in the continuity of a circuit or wire, making it impossible for current to flow.

open (*verb*): To operate a switch so that it breaks the circuit. The opposite of *close*.

overload: Excess current flow greater than the device or equipment is designed for.

panel: A cabinet with its component parts, such as a breaker, fuse panel, or service equipment.

parallel: When electricity has two or more separate paths to travel.

photocell: A device for sensing the absence of light. The internal contacts close to turn on lighting during darkness.

pry-out knockout: A knockout with a slot for insertion of a screwdriver to pry out the knockout.

Romex: A brand name for non-metallic cable used in house wiring.

series: Having only one path for electricity to travel.

series/parallel: A combination of both wiring arrangements.

service drop: Overhead wires from the utility pole to the house.

service head: A special fitting at the top of the service mast or service-entrance cable to keep out rain and point the wires down. The wire entrance faces down at a 45° angle.

service lateral: Underground wires from the utility to the house.

thermal protector: An automatic protective device, usually on or inside a motor, to protect it against burn-out.

Index

Other Bestsellers From TAB